CLIMATE CHANGE

CAPTIVES 2035 & PROJECT SAVE

Carolyn Wilhelm

Illustrated by Pieter Els and Nika Pieterse

Climate Change Captives 2035 and Project Save

Students Help Save the Earth

Carolyn Wilhelm

Copyright

Carolyn Wilhelm, Wise Owl Factory

info@thewiseowlfactory.com

https://www.thewiseowlfactory.com/

Some names and identifying details have been changed to protect the privacy of individuals. This is a work of fiction. Names, characters, places, and incidents are used factiously. Science facts are current for August 2021, and sources are provided in the endnotes.

Book design © 2017, BookDesignTemplates.com

Illustrations by illustrator Pieter Els and Nika Pieterse

http://www.surferkiddies.com
Online cover from Pixabay.com copyright free image

Some images from Pixabay.com, copyright-free image

Some images from Canva.com

Printed in the United States of America

(p) ISBN: 978-0-9997766-8-1
(e) ISBN 978-0-9997766-9-8
US Government Copyright TXu2-156-573

Dedicated to Tristan, Kalli, Crystal, and Phil

MIDWEST BOOK REVIEW

February 2020

Climate Change Captive 2035 and Project SAVE: Students Help Save the Earth

Carolyn Wilhelm's newest Cli-Fi book written for the YA reader is a thought-provoking and engaging read for all ages. Earth has become so polluted people are living in enclosed city centers or are resisters living out in the woods. Students are working to finish the problem solving the adults started. It's a futuristic look at how the world has changed due to global warming with a focus on life in the year 2035. Readers meet the main characters Brea, Robert, and their families and two pen pals Kalli and Tristan from Northern Ireland. All characters are learning to cope in a new world faced with residual effects of climate changes from a previous society who disregarded warnings that greenhouse gas emissions from cars, power plants and other man-made sources would reach a concentration level trapping the sun's warmth near the earth's surface thus affecting the planet's climate system.

I found the book to be well researched with impressive documentation with endnote references at the end of the book. All of Wilhelm's meticulous research on climate change work was woven into the story. Each time Wilhelm used a true climate change fact it was noted within the story text with a little Roman numeral which then references a particular endnote found at the end of the book. In the eBook version, the reader can click on the Roman numeral and the citation pops up on that page. The details in the story are either entirely possible or based on scientific projections of what might happen in the future and how society would cope with consequences.

Characters Brea and Robert's families and Irish penpals are well developed and relatable. Situations each of them face seem entirely plausible with decontamination precautions, polarized sun-protective helmets, limited food sources, growth of dangerous poison ivy plants and limited water supplies. Government control has made laws for families who live within Green City 8763's border. Outside the city limits, severe limits on electricity are placed for those who chose to remain living in single-family homes. Brea and Robert are middle school-aged children encouraged in school to develop research topics for finding climate change solutions. Their efforts become more global with the introduction of Tristan and Kalli, pen pals from Northern Ireland. The years 2019, 2027, 2028, and 2030 are especially important in this story with pivotal climate change events noted. The dystopian type ending is fitting for a few people whose greed and disregard for lack of consequences contributed to their demise as they thought they were escaping the end of life on Earth in a spaceship.

Wilhelm is to be admired for her Call to Action with the publication of this book. She hopes readers will think more about what kind of world they want for future generations and what changes young people can do to promote a more habitable world. It's a valuable resource for the classroom and school libraries. As a teacher, I found the free study guide available at Wise Owl Factory a great resource to use with the book in the classroom.

Sue Ready, **Reviewer**

Free Novel Unit for Climate Change Captives 2035

and Project Save

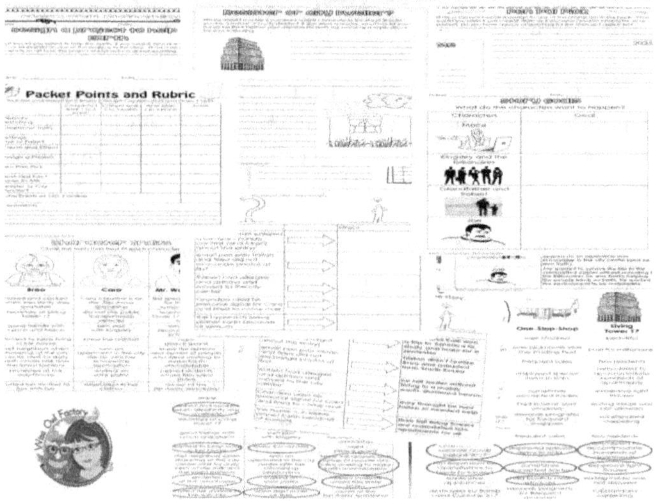

Free book companion novel unit available at the Wise Owl Factory LLC site at this link:

https://www.thewiseowlfactory.com/PDFs/2020/04/book-companion-for-climate-captive-changers-2035-and-project-SAVE

Climate Change Captives 2035 and Project Save is intended for ages 10 and up. The Lexile reading level is 820L.

BREA

June 2035

Brea, June 2035

"Bye! See ya soon!" Brea said as she left study buddy Robert's, her black hair waving in the breeze until she tucked it in the required solar protection partial helmet for her head. She rolled down her long sleeves to help avoid sunburn and slipped on the boots designated for outdoor wear only.

"What starts with an *e* and only has one letter?"

"An envelope. Hopefully ours will have the best news as soon as tonight!"

"Wishing you all the best on your test. We can visit soon! It's summer vacation. See you tomorrow when we know our results," answered Robert. His freckled face was already turning red from being out in the sun a few minutes.

Walking a distance on the absorbent gravel substance safety path, Brea remembered hearing that public transportation had once been available. She, of course, had never ridden in any of the now-outlawed ground vehicles. Roads were generally sunken, broken, and impassable. People now used rocky walking paths, skyways, and an occasional hovercraft. Bicycles were difficult to ride on the wet ground or rock paths.

She wondered about the times before she was ever born. By law, most families now were living in city centers to help control emissions for the failing environment. However, she was going home to one of the only remaining single-family homes just outside Green City 8763's border where she lived with her mother. Living in the house prevented it from being bulldozed, razed, and portions recycled at their expense. Her mother hung on to the home for dear life despite severe limits on the hours of electricity allowed. Having missed a government program to recycle the structure for free, moving was not possible yet. It would be so uncomfortable on this limited-electricity summer day. Anyway, her mum still wasn't ready to move.

Having just visited Robert's earth-sheltered home, she was glad at least for the lack of dampness and mold at home. Earth-sheltered structures had been found empty or built by city center resisters (CCR), so they had somewhere else to live. Resisters were people who did not move to city centers when given their first time-limited opportunity. Peace guards had not bothered arresting resisters because the earth was so soaked. People would move of their own accord at some point, although chances for free opportunities were intermittently offered. Recycling costs were lower for homes built after stricter environmental material laws were enacted.

Minnesota, formerly the land of 10,000 lakes, was becoming one vast wetland. On the walking path toward home, a varmint ran in front of her feet. Brea shuddered. Pest control only took care of animal problems inside city walls. Under the walkway, a drainage tunnel kept the water off the top but attracted many invasive species. Some were easier to deal with than others. She eyed the path suspiciously, glad for fewer insects [i] that might bite. She enjoyed hearing the one singing bird.[ii]

Arriving home, she remembered to put her boots in the decontamination box by the front door. The smell of veggie pizza and the sound of boiling water made her realize her mother was baking and cooking. Not again! Not in this heat! She remembered there were two hours of electricity today, so of course, her mother would use it. Laws prevented air-conditioning use outside the city, and an old fan would provide the only relief. "Make hay while the sun shines," as her grandpa would say. The sun was shining almost all the time, so what did he mean, anyway? Closing the screen door, she remembered to put her polarized sun-protective helmet on a shelf in the hallway. Obsolete and ineffective, sunglasses now were for events such as costume parties. Even movie stars wore the half-bowls on their heads. Not too glamorous.

"Brea, the mail came. I see a letter for you," called Mum. "I think it is your secondary school placement determination results from the standardized tests." Worry gripped Brea's stomach, and she wouldn't open the letter until later when the house had cooled down. The letter could be her ticket to living in the nearby city center. She had applied for a city school scholarship. Resisters had few free opportunities to comply with the new laws. She had checked the box to apply for a grant to pay the house recycling costs. It would be a far reach and only a wild possibility to get even one of the two awards. She held the envelope and put it down. Not yet. Her heart pounded.

Unlike the decades-old K-12 elementary school building provided to resisters; the city center secondary schools were sustainable. Newer buildings offered recycling for everything, including water. Eco clothing and school supplies were now mandated. Bullet-proof glass window walls inside and out were actually solar panels.

Everything was modern and climate-controlled. The letter might mean relief from none, some, or all of the climate challenges she and her mother faced daily.

Dinner was quiet. Brea was hoping the letter would end her school and housing problems. Cara, her mum, was hoping to remain in the house that held her fondest memories. Each said little as the answer had finally arrived. The letter would tell if their hopes were realized or dashed. They washed dishes, tossed scraps of food into the rat-proof compost bin, siphoned water from the rain barrel into the transfer container, and carried the water inside. Some would be used to boil for drinking water the next time electricity was available, and some was intended for household uses. The realization that carrying water might soon not be needed was the motivation Brea needed to open the letter.

In the dark, Brea was more than willing to read it. The solar lamp would not provide enough light for Mum to read over her shoulder. Under the small, soft light with long shadows flickering on the walls, she read her future fortune. She felt like the adult she was becoming, and that she had some control over her life.

Mum was quiet and sitting across the room in her favorite chair. Perfect. First, Brea could digest the information by herself. The only sounds were the tearing of the envelope as she ran her thumb under the sealed edge and the unfolding of the paper. Her extensive efforts to learn were wrapped up on a single results page.

Mother and daughter were each going to be happy in one way for now. Brea had won the school scholarship. She would attend City Center Secondary School 8763. Cara was relieved as she would not have to leave her house. Brea silently shared the news by handing the letter to her mother and went to her bedroom. She fell asleep with her clothes on as she often did, as it saved her from having to carry so much water to wash clothes. She slept knowing she would enjoy climate-controlled temperatures, running water, and hydroponic foods not otherwise available while she attended day classes next year. Her mother would not be angry they had to move.

Alone in the living room, Cara softly cried happy tears. She was not ready to move away. Not yet. She was glad Brea would be spending much of her life at school in more comfort than the old house provided. Relief flooded her body. In complete relaxation, she spent the night sleeping in the chair.

The one fear mother and daughter shared concerned the fact it was challenging to be a resister in a city center. Pressures to recycle their house and live in the city center would mount. They simply could not afford to do so. There were no prospects for having the money to move. None. Brea would have to walk in two different worlds.

ROBERT
2035

Robert, June 2035

A letter had arrived for Robert also, but he was unaware of the delivery. He lived with his grandfather, who could be absent-minded. Grandfather had forgotten to carry in the mail. Morning came, and Robert began searching in the kitchen area for his envelope. He heard Grandfather stirring and patiently waited. He coughed and sneezed as he cleared his stuffy nose, as usual.

For a person with severe allergies, the damp and moldy earth-sheltered house was not an ideal place to live. But his worst fear was poison ivy, which was abundant. Urushiol, the allergenic substance in poison ivy, thrived on carbon dioxide.[iii] Coming in contact with the thriving, spreading plant could cause anaphylaxis, which was not easily treated, especially by the volunteer homeopathic resistance healers.

"Grandfather," Robert asked, "did a letter arrive for me yesterday? The important letter from the school?" Of course, he had to repeat what he had just said to the older man.

"What?" Grandfather asked again.

Robert realized this meant the letter was probably still in the mailbox, and he jumped up and raced outside. Just in time, he pulled back his hand from the poison ivy, which had grown overnight near the mailbox. Grandfather was right behind him.

"Step away slowly, and no one will get hurt," Grandfather joked. Robert, tired of the old joke, followed the direction. Grandfather took a stick and held back the new ivy while retrieving the letter. They both hurried inside.

Using the semi-official resister's Swiss Army jackknife, Grandfather slit the envelope open. He handed the contents to Robert. Standing side by side, both read the information together. Both yelled, smiled, and hugged at the same time. Because of both academic excellence and serious health conditions, Robert and his Grandfather would live rent-free in the City Center. Robert would attend secondary school there, they would share an apartment, and finally, the earth-sheltered home would be destroyed and recycled by the government at no cost to them.

"Surely, Brea's letter will say the same thing! Her grades were often better than mine!" Robert hoped. Saying he would be back for breakfast; he walked as fast as he could all the way to Brea's house. Wheezing and coughing, he knocked on the door with a huge smile that said it all. Brea's face fell when she opened the door, so that -- for a second -- he didn't think she won either award. Brea stepped outside and shared her news.

"At least we will both be in city schools," Robert realized. He was still radiating total happiness. "You will be in City Center Secondary School, 8763, right?"

"Yes, the only secondary school allowing students from outside the city," Brea confirmed.

"I'll be attending Secondary School 17, the one for our living tower."

"Good to know, Robert," said Brea, smiling. "We will have joint symposiums and see each other, sometimes, dontcha know."

It was too bad the invasive species city boundary legislation prevented much interaction between those who lived beyond the city and those who lived inside the borders. It took too much time and effort to decontaminate from invasive seeds and other pollution to allow students from inside and out to attend the same schools.

Brea was happy about her schooling award, although she had wanted both awards. Robert suggested, "There is another chance for you to get a new apartment, you know. Invent something to help solve climate change problems!"

That just sounded like pie in the sky to Brea, who congratulated Robert before going back inside. She was happy for Robert with his health problems. And his house was less costly to recycle, so the award was more easily earned. Still, she had been too hopeful. Well, she would just have to solve climate change, no biggie, sure. She could start on it now, not that governments, countries, corporations, scientists, and committees had only made some progress in coping with the effects. Yeah, she could probably solve everyone else's problems.

Brea had to keep in mind her mum was happy, and was truly thankful she herself would be in comfort much of most days. She could now only visit Robert's new home on symposium days. She earnestly hoped the new building conditions would help improve his health.

GRANDPA
June 2035

News by Night

Near dusk, Brea noticed the solar communicator was flashing Grandpa's unique signal. What was wrong with Grandpa? Mum stood frozen near the communicator. Glancing at each other and without talking, the pair grabbed their overnight bags and began race walking to Grandpa's house, thankful for the solar lighted gravel paths. Each time he was sick, he became frailer. He had never made sense of having to boil his drinking water, and so occasionally, he just would not do so. He said he was a resister and was resisting boiling water. When the city centers were under construction, he had found a tiny house just outside the city limits. He squatted in it and had never been questioned about ownership. It was his happy place when he felt good. It was a danger when he was ill because of thick mud and puddles in the yard. He had previously slipped there when gathering scarce clear rain barrel water. Broken bones were his specialty. Brea and Cara helped around the house by bringing in the water and taking care of things when they could.

Brea felt a little guilty. Grandpa could have lived in the city with Brea and her mother. If only. Maybe she should have studied more, she thought, but also realized she had given it her best. During the school year, she studied until exhausted each day. There was no more she could have done.

Arriving at Grandpa's, lights appeared in the darkness one by one until it was obvious something was happening. Solar lights in the resister area were not strong since replacement parts were difficult to get outside the city. If everyone was using the precious lights, it meant there was news or some emergency. First, they had to reach Grandpa. Arriving, surprise! He looked fine and was standing on his steps, holding up one light. Putting a finger to his lips, they remained silent. Quietly people gathered, ready to hear the news. Waving lights meant good news. Still lights meant bad news. The lights were still. Brea wanted nothing to ruin her new opportunity now!

Grandpa spoke. People who were close enough to hear listened, and those who were too far away waited for the message to be repeated to them in a few minutes. There could be no shouting in the woods to alert night peace guards who passed information on to the city police, even if they had stopped enforcing some laws.

"We have news. Please do not shout or speak. Please discuss with your own families. First, several of our students will attend City Center Secondary School 8763. We will need our recent graduates to hand down their electronic study libraries to the incoming students as soon as possible, so they may begin the required reading before classes began."

"Will you give us a list of students?" asked Mrs. Montes.

"Yes, check your messages later tonight. And I have more information. Another bank has sunk, and divers got the vault open. They believe no one followed or observed the activity. The designated person in each family or group of friends will share and advise, yeah."

"Don't forget to mention the epi-pens," suggested Colton, a volunteer paramedic.

"Right, the emergency store of insulin and epi-pens is dangerously low if anyone can help. *If asked, that was the reason for this group communication.*"

"Anything else we need to know?"

"We need to disperse the funds immediately. Take this seriously. You may quietly leave."

The crowd whispered the message to those who could not hear as the people returned to their homes. A five-year-old squealed with joy when learning about the money and was quickly hushed. Centrally located money would soon disappear; quick distribution was essential.

"Why didn't you use the happy solar signal and wave the lights?" Brea asked once inside the tiny house. "Why did you scare us at home by using your personal distress signal?"

"Happy people create more curiosity among the peace guards," responded Grandpa. "Larger sums of money have to be dealt with rapidly, eh?"

"Is the money enough to afford our house recycling?" hoped Brea, glancing at her mum, afraid she had offended her.

"No, nothing like that." Cara breathed a sigh of relief. "But we might have more food at home for about six months."

Cara mentioned Brea's scholarship, and Grandpa was quite proud she had won. He added he hoped to move to a living tower someday, but not yet.

"Staying overnight?"

Mum answered they would stay over as it was pitch dark now. The tiny house was built in 2015 when such houses were the latest craze. The thought at the time was that people could spend their money on travel and experiences rather than having a home mortgage. Comfort wasn't the major feature, and it was twenty years old. There would be much to think about if they could not sleep well on the wooden floor.

MORNING
AT GRANDPA'S
Tiny House 2035

Morning at Grandpa's Tiny House, 2035

Brea was up first. She fished through her overnight bag for the school communicator device. She sent a message to Robert using a secret code they had devised years ago. Several students in their class had developed secret languages. It wasn't difficult, and grown-ups didn't understand it. She wrote, hoping Robert was awake.

Brea to Robert (translated): DID U HEAR?

Ding. Brea turned the volume down and listened to sounds in the house. The beep had woken no one. She waited and almost fell back asleep. Noticing the flashing screen, she picked the device back up.

Robert to Brea (translated): BUYING FOOD WILL BE Gr8! REAL STORE!

Brea to Robert (translated): LET'S SHOP TOGETHER. BYE FOR NOW.

The friends each turned their communicators off as they had few chances long enough to recharge their slow, old batteries. A little messaging had to go a long way.

Food had evolved since they had been born. There used to be foods such as apples, corn, wheat, and potatoes. Corn was unavailable as the temperature had warmed so much that crop failed. Corn was in almost every food known to man in a short ten years ago. Wheat yields had fallen so much that Mum could not bake bread and cake. People who baked now made mostly "original and unique" green creations if alternative flours were not available. Formerly, there had been holidays with special meals such as something called Thanksgiving.[iv] Now it was everyone for themselves, even growing plants indoors, finding old cans of rations, or the ever-popular elementary school breakfasts for all families. The school sent home a few canned or boxed provisions each Friday to most families, although it was never anything delicious. Food kept people alive. There was not much variety, and it was rarely used for entertainment or parties.

To go to an actual store and buy food and possibly milk would be a treat. Everyone couldn't rush to the store immediately as it would raise suspicion. Waiting for a turn to shop would take patience. Leaves were easy to grow but did not fully satisfy at mealtime. Scientists knew even way back in the year 2015 that heat-stressed pigs had about a fourth less protein than those kept in a cooler environment.[v] Brea wondered for the millionth time why climate change had not stopped, even with the drastic measures implemented worldwide. What was wrong with grownups? Not that she wanted to eat pork; it was just a fact that had stuck in her mind. She felt it might also apply to human muscles and shivered at the thought.

PICKING
Poison Ivy

Picking Poison, 2035

It was Ivy Duty Day for Brea and Cara. Not fun, but so necessary. Wearing high wader boots, protective clothing, hats, and long gloves, they began pulling the weeds wherever they saw them. At least they didn't have to wash the special clothing themselves, as it would be too dangerous. They would work for two hours only, rinse with special soap from an outdoor shower, remove the volunteer uniforms, as well as hang them on branches on a designated tree. They would return home with wet but safe clothing where they would towel dry. The warm, dry air would take care of the rest.

"Oh, a loose solar rock. Finders, keepers," said Mum.

"I'm going to find more than you today!" challenged Brea.

The little game helped make the task bearable, and their collection was growing. Someday it might prove helpful.

They knew burning poison ivy would release urushiol and possibly create a life-threatening reaction in the lungs.[vi] There was no fast way to deal with the plants, as it was dangerous to burn them.

"When I was young," Mum said, "if we touched poison ivy, we washed with Fels Naptha soap right away. That stopped the rash from spreading. If we forgot, were we sorry!"

Brea wondered if the protective clothing was washed with that soap, but she already knew her mother didn't know what they used. And now it probably took more than just that soap because poison ivy had become so potent.

"Remember, mum, the botanist who rolled around in four leaf clovers and poison ivy?"

"It did not give her a rash of good luck, and it wasn't me," answered Cara. Joking helped them get through the task.

"Mum, later you can relax in your rocking chair!"

"Groan!"

"What did the scarecrow resistance worker say?"

Mum laughed and pretended to think hard. "Oh, let me think. This is difficult! Hmmm, I know, it is in my jeans."

Working side by side, Brea asked her mother about the City Center Resisters. Resistance fighters in stories she had read differed from the ones today. "Why are they called resistance fighters? They rarely fight," questioned Brea.

"They fight the effects of climate change," responded her mum. *Cara thought to herself that volunteer groups gather to fight invasive species, provide whatever healthcare possible, share money, and divide resources equally. Everyone's energies went to survival. Rarely did any conditions improve.* "We are working as resistance fighters right now," she added. "At first, it simply meant people who didn't jump at the chance to live in the city centers, leaving almost everything behind. I just wasn't ready at that point in time."

Brea realized resister efforts went toward trying to survive beyond the city limits as they helped and supported each other. It just wasn't how she pictured fighters. She had long realized her mother might never want to leave her precious house.

"When were they started? Whoever thought of them?" asked Brea, tugging up two intertwined plants at once.

"As far back as 2017," started Mum.

"That's longer than I have been alive!" interrupted Brea.

"As I was saying, in 2017, the American president asked for permits in Ireland to build 14-foot-high walls around his golf courses and resorts while denying there was any climate change," continued Cara.[vii]

"And you know about that because of your Irish heritage?"

"Well, partly, I suppose that made me pay attention. From about that time, rich people were building walls around resorts and luxury hotels. They added drainage under the areas."

"Yah, it runs out near our house and goes down the hill! But weren't some areas super dry and without water?"

"Right, wetter, drier, hotter, and colder. Climates became more extreme."

"Ouch! So, people moved?"

"Yes, if they could. The CCR (City Center Resisters) didn't resist the City Center (CC); they could not afford to live there. It had been an issue of the time to gather family members to move immediately, which was often not possible. Now it requires economic help and retraining to fit in or work in city communities. The economic ladder outside the centers was rungless. Small farming was no longer possible due to flooding. Financial issues outside the city borders were virtually incapacitating. Constant humidity left people with little energy. They were just left out."

"Do you think people in the centers have more energy and better health?" asked Brea. Looking in the glass windows visible from outside CC 17, it seemed there was so much movement and activity, anyway. Who knew for sure? Students who attended City Center Secondary School 8763 only learned the school curriculum. There were no opportunities to explore the city. School. Home. School. It did give a person pause. Information about life in the city was not widely known outside the boundaries.

"No one really knows," Mum answered. "Maybe you'll find out more next year at your new school," she hoped.

"What is being done with the ivy we collect?" wondered Brea.

Mum answered again that no one really knew. They both understood it could not be burned or eaten.[viii]

CITY
CENTERS
Housing for All

About City Centers

Robert learned similar information from his grandfather. "City Centers? How did that happen in the space of a few years?" Grandfather thought. "Oh, the people in power knew what they were doing by ignoring climate change warnings and increasing many emissions and carbon dioxide for profits. They knew because they had already been storm-proofing luxury properties as they prepared for the eventual demise of conditions for life on Earth. They privately acknowledged climate change because they tried to protect themselves. Early in the 2000s, some businesses and governments were being deceitful."

"Why didn't they hurry to help clean up the air and environment?" Robert wanted to know.

Grandfather explained the expectation was that by 2050, there would be disastrous climate change issues. He said, "Maybe that seemed like forever to the greedy. Maybe it seemed like enough time. Many people did not think the change was imminent. Some people did not believe in science facts then. In 2014, the Pentagon, however, had described climate change as an urgent and growing threat to national security. [ix] Climate change migrants who walked or tried using boats, both of which could be personally dangerous."

Grandfather continued. "Tables turned. The rate of climate change increased as regulations eased. Coastal erosion rates more than doubled.[x] Larger, more frequent storms occurred. The protected areas did indeed survive better than outlying areas. Specifically, many protected areas survived nicely. Class-action lawsuits about 2028 gave the public a right to live in the areas the rich thought would protect only the top one percent financially. Now, most people were living in areas the wealthy thought they had reserved only for themselves. It was a small satisfaction for those who lived through storms, earthquakes, disasters, the pandemic, and loss of family and friends, only to have to adapt to a new way of life. They felt like foreigners in their own countries. Still do."

"Construction of the cities was rapid, right, Grandfather? And they just stopped repairing roads, communication systems, and infrastructure?"

"Right, there were lawsuits once the population started being devastated by disaster and illness. Many non-essential businesses closed, and everyone had to work on adapting and building living towers. It was the only way to save as many lives as possible. It was clear humans would suffer too much from climate change." Grandpa didn't want to continue the conversation. He had explained this before.

"Why couldn't you get to the City Center, Grandfather?" Robert asked for about the hundredth time.

"Well, you can tell me by now," Grandfather prompted.

"You were worried about my parents?"

"Yes, and you know the rest. I thought they could find us easier if we lived near the city. Remember, this earth-sheltered home was just a place to enjoy in the summer and only live in if worse came to worst? And it did."

"We have each other," said Robert. "Are you sure my-parents' house was hit by strong wind and hail, and then destroyed in a mudslide on their hill during flooding?"

Grandfather looked at Robert, kindly, and hesitated. Flooding had been going on for decades. "Well, the house is gone, that's true. Friends have verified the neighborhood is gone."

The pair quit talking. They hadn't heard from Robert's parents for several years. Discussing possibilities and wishes was pointless.

REQUIRED
SUMMER READING
2035

Lorem ipsum dolor sit amet, consectetuer adipiscing elit, sed diam nonummy nibh euismod tincidunt ut laoreet dolore magna aliquam erat volutpat. Ut wisi enim ad minim veniam, quis nostrud exerci tation ullamcorper suscipit lobortis nisl ut aliquip ex ea commodo consequat.

Duis autem vel eum iriure dolor in hendrerit in vulputate velit esse molestie consequat, vel illum dolore eu feugiat nulla facilisis at vero eros et accumsan et iusto odio dignissim qui blandit praesent luptatum zzril delenit augue duis dolore te feugait nulla facilisi.

Lorem ipsum dolor sit amet, cons ectetuer adipiscing elit, sed diam.

Lorem ipsum dolor sit amet, consectetuer adipiscing elit, sed diam nonummy nibh euismod tincidunt ut laoreet dolore magna aliquam erat volutpat.

Ut wisi enim ad minim veniam, quis nostrud exerci tation ullamcorper suscipit lobortis nisl ut aliquip ex ea commodo consequat. Duis autem vel eum iriure dolor in hendrerit in vulputate velit esse molestie consequat, vel illum dolore eu feugiat nulla facilisis at vero eros et accumsan et iusto odio dignissim qui blandit praesent luptatum zzril delenit augue duis dolore te feugait nulla facilisi.

Lorem ipsum dolor sit amet, cons ectetuer adipiscing elit, sed diam.

Lorem ipsum dolor sit amet, consectetuer adipiscing elit, sed diam nonummy nibh euismod tincidunt ut laoreet dolore magna

Required Summer Reading 2035

Brea was reading her electronic library as part of the summer education preparation for school. In 2028, vehicles near passable roads had to be driven to recycling centers, crushed, shredded, separated into small pieces, and sorted into various metals. The largest such shredder in the world in 2016 had an 8000-horsepower electric motor to crush six pre-crushed cars per minute into fist-sized metal chunks.[xi] One built during 2030 was about ten times as fast. Worldwide, no one wanted to give up their vehicles, trains, buses, and planes. An international court decided, and there were no exceptions.

She liked the international court. She liked it that because war was detrimental to the environment, was also outlawed. Some illegal fighting continued. She considered the solar rocks she and her mum had picked up along with poison ivy. They were legal to pick up and take home only if found in mud and away from well-engineered solar trails. She worried about being caught with them, anyway. They had dozens hidden in the back of a cupboard. It was just they had so many. Who would believe their rocks were not taken directly from the paths? Everyone gathered anything that could be useful someday. Who knew which items were pointless to collect and which might be helpful? And who knew which might cause trouble with the peace guards?

Next, she read an article from 2019, which stated Minnesota had begun its descent to an unrecognizable future[xii] NOAA had ranked Minneapolis and Mankato as the second and third fastest-warming cities in the United States.[xiii] Way before she was born, changes could have been enacted. Lakes had been freezing later in the year while thawing earlier every spring. Changes in growing periods and voracious new pests had been noted. The famous loon migrating waterbird had left the state.[xiv]

Enough reading. She could only take so much of this type of information at a time. Next time she would read about the new hopeful green inventions and companies now trying to fix climate-change problems.

SUMMER SOLSTICE

Burning Gloomies

Summer Solstice, 2035

The summer solstice evening was fast approaching. Although in name and date only now that the seasons had changed so much, it would be a fun event. She couldn't imagine a time when summer was only a few months long. In her mind, it was the season with the most apparent climate change impact. Winter still had a variety of precipitation such as ice storms and strong winds, but summer weather seemed more volatile to her. She knew long ago meteorologists were prevented from discussing climate change to help keep ratings high.[xv] Ratings, when lives had been at stake. Right.

She was looking forward to the one bonfire of the year. Due to the fact oil and gas sometimes seeped up from the wet earth, fires were severely regulated. People would be throwing their "gloomies" into the flames. Gloomies were handwritten notes about things people would rather forget, things they wanted to change in the coming year, things they were thankful for, and other thoughts they wanted burned in the fire.[xvi] The event helped everyone have a fresh start and a sense of relief. Firefighters would be on hand to quickly put out any possible fires. Brea had read about fireworks in her history class but would never see any herself.

"Will you be there tonight?" asked Brea when she visited Robert, who was already beginning to pack.

"Wouldn't miss it for the world, of course."

"I'm glad you haven't moved yet, at least we can enjoy one last summer solstice celebration," said Brea. There was a little sadness in her voice.

"We can text," remembered Robert.

It was true, but it just wouldn't be the same. Before monthly joint symposiums, students from outside the walled area had to go through a lengthy decontamination process than the usual daily one at their own school before walking over to Tower 17. This was to help limit spreading invasive species inside the city. What a pain. The schools provided disposable, recyclable, sustainable clothing and shoes for these events, at least. The showers would not be so bad, and at least they were indoors.

After dark, the fire was started and grew to be huge, rising to the sky and sending sparks back to Earth. The smell of burning branches filled the air. Considered a symbol for survival, it offered all who attended a fresh start for the rest of the year.

It was time. People gathered around the fire, greeting one another at the allowed event.

"Today is the summer solstice...I have a feeling you all have had a long day," said Grandpa when everyone gathered. There was a scattering of laughter as people began to listen. "For the past couple of years, I have been saying that the only holidays worth celebrating are the equinoxes and the solstices. I find all the others to be astronomically unimportant." More laughter, he was warming up the crowd. Suddenly serious, he asked, "Have your Gloomies ready?"

"Yes!" people answered, quieting down.

Quietly and one by one, people put their private, folded writings into the fire. Sadness and guilt dissolved into ash. Hopes and wishes for the rest of the year began afresh.

Robert and Brea had several Gloomies their friends in Northern Ireland sent to them, as well. Written months earlier due to the slow postal system, the envelope arrived in time for the thoughts of their friends to be added to the fire. Because Gloomies were not shared, no one realized if a few extra pieces of paper were being burned. Their friends joined the event in spirit. The same online texting friends who were working on a startling idea to reduce carbon dioxide. Could the pen pals help solve some of the climate change issues? Kids? Robert and Brea thought maybe so. As far as they knew, the teachers didn't realize the scientific depth of the letter-writing information . . . yet.

Traditionally, a fire should be allowed to burn out naturally, which was no longer possible under the law. Oh, well, the best part was still allowed even if they could not watch the glowing embers.

The students entertained those gathered with short skits, stories, and poems while the fire still burned. Funny skits, sad stories, and thoughtful poems helped celebrate the event.

Robert's reading of *We Grow Accustomed to the Dark* by Emily Dickinson [xvii] earned a standing ovation.

The poem reflected the feelings many had about surviving the seemingly unsurvivable times. They had all lost so much. Persistence and patience had helped them through the changes.

More hours of sunlight each summer day would cause more difficulties. Hope for a better future was on the minds of most people.

By Emily Dickinson

We grow accustomed to the Dark --
When light is put away --
As when the Neighbor holds the Lamp
To witness her Goodbye --

A Moment -- We uncertain step
For newness of the night --
Then -- fit our Vision to the Dark --
And meet the Road -- erect --

And so of larger -- Darkness --
Those Evenings of the Brain --
When not a Moon disclose a sign --
Or Star -- come out -- within --

The Bravest -- grope a little --
And sometimes hit a Tree
Directly in the Forehead --
But as they learn to see --

Either the Darkness alters --
Or something in the sight
Adjusts itself to Midnight --
And Life steps almost straight.

ROBERT'S

Moving Day

Robert's Moving Day, 2035

Robert and Grandfather's change of address happened quickly. Once the government decided to destroy a home, there was never much warning.

Brea was helping Robert take down his posters and roll them up. The images and quotes by Greta Thunberg, Jane Goodall, and Zero Hour teens had seen better days. He had already packed the precious few photos of his parents.

"Do you think they will let you keep all the posters?"

"Yeah, for now, they might not all pass mold inspection, but I can try," Robert sighed.

He hoped their belongings would be dry, not damp all the time in the new place. He packed his chess set, textbooks, assorted games, music device, electronic school reader, school texting tablet, and clothes. Essentials, in his mind. Robert and Brea put them in packing boxes which would be picked up by workers. Of course, everything would have to go through the decontamination process, which would require a day or two. Late in the afternoon, they would take decontamination showers, change into government-issued temporary clothing, and find their new place in Tower 17.

"The movers will be here soon, yeah," called Grandfather.

"I know, I'm almost finished," answered Robert. "Do we tape the boxes shut?"

"No, they have to check the contents, and then do the taping."

"OK, five more minutes!"

Brea and Robert had some time to go shopping at the store before going to her house. They couldn't wait. It was a chance for some foods they rarely ate "I want some chocolate, and maybe beet or cassava chips," stated Brea.

"That would be fun," Robert agreed. "Those would be good! Do you have your money?"

"I can spend what's on my card," Brea informed him. Brea's mother had thought cash would be suspicious, so they had added it to her student account. She would use her debit card. "One of us with cash will not arouse so much suspicion!"

THE STORE

Used and New Items

One Stop Shop

Robert's last afternoon outside the city walls was spent shopping at the One-Stop-Shop with Brea. They were more than ready. It wasn't like the store was anything special. It had a captive group of resister shoppers and ran no loyalty or rewards programs. None were needed. No shoppers were "just looking." Sales and discounts were rare. Foot travelers who happened upon it might try to sell or trade a few items so they could continue hiking long distances. One of the store's nicknames was The Trading Post.

"We will look at what is in stock and decide if we want an item or not," said Brea.

"Right," answered Robert, "and we should be careful with our money."

"Oh, I will be. I'm going to keep a running total in my head. I can't overspend."

When they opened the doors, hot air greeted them. The store used to be a beauty parlor and had fixed windows which could not be opened. The large glass windows were covered with insulation and scraps of wood to help the rooms retain heat in winter. It seemed odd even to infrequent shoppers to see hair-washing sinks full of soup cans or shoes. Special items were placed in front of large mirrors, grabbing attention to help increase sales. One nice part of the shopping experience was the lovely smell of freshly cleaned surfaces, although probably scrubbed with now-banned chemicals.

The second floor was up the escalator which was used a staircase, the former location of the plastic surgeon's beauty Botox services.[xviii] Of course, the escalator had been turned off to reduce consumption of electricity years ago. It was odd to think people used to ride up and down instead of climbing stairs. Conveniences such as that had contributed to climate change.

Near the door was the ever-depressing bulletin board with photos of missing people. Robert noticed the now-faded picture of his parents.

"Try not to think about them today," Brea said softly. "This is an important day in your life, moving to the city center. I know they would be so proud of you."

Robert turned his face away and blinked back a tear. "Thanks, let's start upstairs," Robert suggested. Games and art supplies were on the second floor.

"Let's only look upstairs first, and then shop downstairs. We can go back up after we find some of the food and used necessities on our lists downstairs. Essentials first, then games, crafts, and new items. I am glad we don't need new helmets as the price has gone up so much."

"Agreed," Robert said as he started taking two steps at a time. Brea was right behind him.

The two students did not comment much about the games, crafts, electronics, and cool items on second floor. They caught each other's eyes a few times as they admired things they might want or didn't realize existed.

Back down on the first floor, they each took a basket and started finding items on their lists. Available essentials should also be considered. They knew not to talk too much and remained businesslike.

Brea's mother was hoping to have milk or cream for tea. Brea found a quart of milk, although it was only 1% and not exactly what her mother wanted. She put it in her basket as it was the only option. She started looking for thread, needles, and detergent for washing clothes.

Robert's list was not very long and was flexible. The items on it included canned soup, canned vegetables, a school backpack, and coffee. Robert was able to find the tomato soup he liked, and had to pick some vegetables he didn't to add to his basket. He found a two-pound can of coffee beans, also. Grandfather would be very pleased. Coffee was a rare treat. He would carry the items in his backpack to the new place for food if the living tower cafeteria was ever closed. By getting mostly things packed in cans, he knew it would be easier to enter the walled area tonight.

When their lists were as complete as possible given the limited stock available, the two friends again visited the second floor. *Brea wondered why the fun and entertaining items were so expensive.*

Robert decided on a second tablet for his Grandfather since he would have his at school so often. They could then text each other. He also found some drawing pencils for himself. There was a new backpack, but he had opted for a used sustainable one downstairs. That was all he could afford, and anyway, everything would have to go through the entry process as the boxes were now packed.

Brea was able to find a watercolor paint set and a few pieces of watercolor paper. She and her mother could paint and perhaps use each side of the paper.

On the way downstairs, they saw a few neighbors who had not waited for their turn to shop, so they quickly paid and left. It was important to keep to the shopping schedule to avoid suspicion about the money from the bank vault. Well, at least they had adhered to the schedule, were first to shop that day, and didn't converse with the neighbors. They had followed the rules.

Outside, they joked about the shopping experience. "Nice delicatessen, right?" asked Robert.

"Quite the jewelry store!" said Brea. "I'll return to that salon, soon, for a manicure!"

"Well, I liked the bookstore best, yeah!" laughed Robert.

Brea suddenly felt serious and asked, "Do you think there will be nice stores in the walled city center?"

"Hey, we will need a code to discuss things like that, as I don't know what I can share or not once I'm there," Robert said thoughtfully.

"We have a secret code already, remember? But we will still have to be careful," answered Brea.

The friends walked home together one last time and spent the last few hours at Brea's with her mother and Robert's grandfather. Then, they got the message from the living tower. The government workers had finished moving Robert and his Grandfather to their new place. Then it was time to leave.

"It's been a great last day. I hope your new apartment is very nice. I'll look forward to the symposiums when I can see you again."

"Thanks, and I'll miss you, study buddy!" said Robert.

TRISTAN

Norther Ireland

Tristan, Northern Ireland, 2035

Tristan lived in Northern Ireland. He was a serious homeschool student. His hobbies included nanoengineering and 3D graphic design. Of course, he spent time reading and taking extra STEM classes. Twelve years old with brown hair and eyes, he had earned a black belt in Ju-Jitsu. He could take care of himself. He would roam the woods near his home, promising his parents to stay away from the cliff and local coastal caves near the Irish Sea. Sometimes his whole family went on a dander with him. Norn Iron's term for a walk was dander.

Tristan always found it interesting to learn about places like the bog near Ballycastle, which has many rare plants and animals. A fact that especially fascinated him was bogs do not require much oxygen. He read about them on the NHPBS website and learned:

> *"Bogs* have low levels of *oxygen* in them because water doesn't flow in and out of them easily. Low levels of *oxygen* and cold

temperatures make it more difficult for fungi and bacteria to decompose dead plants quickly. This helps peat form."[xix]

He realized the harsh environment might somehow play a role in helping climate change problems. He knew bogs have one-third of the world's organic soil carbon, which plays a vital role in mitigating climate change and stabilizing the carbon cycle. It was something to consider.

"What about forests, eh?" his sister once asked. She was always talking about trees.

"Bogs have twice as much organic soil carbon as standing forests," [xx] he had knowledgably replied. "But promoting proper maintenance of forests should also be done as poorly maintained forests do not give off enough oxygen."[xxi]

"Are you going to try to find dead bodies in the bog?"

"No, this is nothing about finding bog bodies. Just because bogs prevent decay and mummify human flesh causes some people to be interested in the bodies. I am thinking along a different line entirely."

"What?" his sister pressed.

"I'm thinking of entering a climate change contest," he admitted.

"Oh, one of those again," sighed Kalli.

Still, he wanted to learn more. Some tropical peatlands are hidden. No one knows where they are. They should be preserved. It takes money to test the soil and find where this valuable, helpful type of land is hiding. He had an idea that he would keep to himself for now. He had so many ideas and needed to make some decisions as they couldn't all be promoted.

Through the school pen pal program, Tristan became friends with Brea. He selected her name from a list the teacher shared because Brea's name sounded Irish. It was Brea's mum who was Irish, though. Her parents had found her through an adoption agency. Brea thought it was fun Tristan assumed she was Irish and let him continue to believe so as she thought they would never meet. She was actually born in South-Korea when international adoptions were still possible.

Tristan and Kalli had not divulged to Brea and Robert they were, in fact, brother and sister. It was more interesting to correspond as if from different households. They would laugh hilariously when they received messages which indicated their pen pals had no clue they were related. For instance, their pen pals had written it was such a coincidence they were both learning at home. Sometimes Tristan and Kalli had inadvertently both used Norn Iron spake (Northern Ireland wording), such as using the word "poke" (ice cream) on the same day. Brea noticed.

"Do you know Kalli?" Brea wrote as she was becoming a little suspicious the two knew each other. She mentioned her friend Robert was pen pals with a Kalli who also lived in Northern Ireland. She was starting to think they might be cousins or siblings. Their little secret was fun, and not in a million years did they think they would ever visit America or be anywhere near Green City 8763 until a certified letter arrived.

"The winners of the trip to America for the International Secondary School semester are Tristan and Kalli," the letter announced. Their application to the secondary school climate change contest had been selected. They had long forgotten about their contest entry.

"We won! And we won't even be old enough to attend high school for another two years!" exclaimed Kalli.

"The wee ones are geniuses," said their father. "Their mother is a brilliant homeschool teacher."

"You help us learn sometimes, too, Da," replied Tristan.

Lo and behold, the four pen pals would indeed be meeting.

"You can stop entering contests now, Tristan," Kalli giggled.

Now that we are going, I wonder what my individual project should be, thought Tristan. *There are so many possibilities, which is a good thing, but I should pick one of the ones we wrote about on our applications to visit Minnesota.*

Tristan was considering several ideas such as the growing weeds if nothing else was possible, protecting peat bogs and seagrass, improving forest management, and improving farming methods could all help remove carbon from the atmosphere. He needed to focus on one idea. He was considering planting flowers as natural insecticides such as marigolds and nasturtiums,[xxii] but there were so few insects now, anyway.

He had read, "The beauty of weeds is that they also act like a carbon sink: a system that takes carbon out of the atmosphere and puts it into another form of storage." Now he didn't make fun of people who had weeds in their yards. Enough people growing weeds would make a giant weed sponge! Whatever would help![xxiii]

He wondered where he could make the biggest impact.

And, of course, only he, Brea, Kalli, and Robert knew about the secret group project, Project SAVE.

Soon, it would be time to travel.

Privately, Tristan thought he wasn't really sure he wanted to go to America. They used to have so many school shootings there. He remembered it was also that one of the main countries that created so many problems for the environment. Of course, he did not say this out loud to his sister or parents. The laws had been changed, and no one was supposed to own a gun except in hunting clubs. But he knew there were many resisters living in the woods. He wondered if they were all adhering to the laws. Robert and Brea were resisters who did follow the laws, but who knew what might happen? If he stayed in Northern Ireland, he could finish a nanoengineering course and start another 3D design course he really wanted to take. The design course might not be offered again for a few years. He would have to take an incomplete for the engineering course in progress, finishing it upon return. He worried about the trip, knowing he would go, of course.

KALLISTA
Northern Ireland

Kalli Northern Ireland 2035

Kalli was writing to her pen pal, Robert. Eleven years old with a bubbly personality, she had just returned from walking around the neighborhood selling fresh-baked muffins and cookies. With brown hair and eyes like her brother, she was often demonstrating one or another of her mum's curriculum reviews on YouTube for other homeschools. Her mum had a never-ending stream of lessons other people wanted her to review. Kalli could make videos in her sleep anymore and had learned so many interesting things.

There used to be a delivery van which would stop by to drop off the numerous packages for her mother's blogging business. With the infrastructure practically ruined due to worldwide flooding, deliveries were no longer possible. Now the family walked to pick up packages and carried them home.

She asked Robert about his new place. She knew it would be good if he could live in healthier conditions, and was glad he had won the award for an apartment in a living tower. She was glad he could now study in a climate-controlled school.

"Do you think you will feel better there?" she wrote.

Robert wrote back describing the amazing difference he felt already. Outside the CC, he had been tired and couldn't walk very far. He wrote he had thought it was due to having asthma and allergies. He thought his Grandfather was not very energetic due to his advanced age. When they moved, they discovered enough indoor oxygen made a dramatic difference. Grandfather was livelier and said he even felt younger. Robert felt he had more strength and energy than before, although he still had some asthma and allergies. Outdoors, oxygen depletion still affected both of them.

She was surprised to learn although he had packed his moldy posters from his bedroom wall, the government had had them reprinted and they were fresh and lovely. They had replaced some of his clothing considered too contaminated with identical new clothing so he would feel at home. Even his "new" used backpack from the One-Stop-Shop had been duplicated. He could tell as it was brand new.

"Where do you live? I know in Northern Ireland, but do you live in a tower with oxygen, or what?" Robert wrote.

"I hear we will be meeting at the climate change symposium at your school, so I may as well confess," Kalli wrote back. "I live in a house with my brother, Tristan."

"Tristan? Is that the same one Brea writes to?" asked Robert. He hadn't believed Brea's suspicions the two were related.

"One and the same," she admitted.

"Oh, she is my friend, and we had no idea you two even knew each other," he replied. "Ha-ha, you must have had some fun with our messages."

"Well, we were sharing our climate change studies, and it was best to keep the information separate," we thought, Kalli said. "We have to meet as we have ideas we want to run past you and Brea!"

"I bet they will be outstanding," answered Robert, adding he needed to stop and charge his tablet.

"Bye!"
"Bye!"

Offline again, Kalli wondered and asked Tristan, "How will we travel to Minnesota?" The roads and highways had not been repaired for many years.

Personal vehicles were prohibited. Cruise ship trips were few and far between. Airplanes trips were discouraged. Travel was difficult. She used Norn Iron spake when she said she was *"ascared."*

"What? What do you mean, how will we travel?" answered Tristan. "Airplane to Minnesota, and then hovercraft to Green City 8763's accommodations and school."

"Are we going to have to be decontaminated?" Kalli was still questioning the process and plan to travel to and from Minnesota. She would be turning twelve once there and thought about having a birthday away from home.

"Yes, we will."

Having to be decontaminated along with all their clothing and travel items sounded depressing to Kalli. She remembered Robert and his Grandfather had to do that to move, and Brea was going through that on school days by living outside the city walls. Still, it was an unwelcome thought. They were lucky enough to live on high, dry land and did not have to join a walled area. They felt free and unrestrained, indoors and outside.

"Why do we get to fly, again?"

"We are working on solutions to help restore the carbon balance, so we get an exemption and can fly. We may not get to fly again in our lifetimes," explained Tristan.

"But we get to fly home to Northern Ireland, right?" she asked again.

"Yes, right."

"But not our parents? They can't come?"

"No, they can't. It will work out to be fine. Let me study now, Kalli."

Kalli knew she could rely on the secret project the four pen pals were working on, but she wanted an individual project as well. She also wondered what to pick. There were many possibilities.

She didn't want to tell her brother, but she was concerned about the trip to America. She loved baking cookies and selling them to the neighbors. She had heard there would be little of any kind of flour in City Center 8763. She did know how to bake with several kinds of flour. She kept pestering her brother with questions.

"How will I bake when I am in Minnesota if they don't have something I can use? I can bake with chickpea flour, arrowroot, cauliflower, or rice flour."

"Sis, the school is going to feed us so you won't have to bake. Don't worry about cookies."

"But what if I want to bake?"

"We will be too busy with our science projects and presentations. As I said, don't worry about baking."

She was beginning to email her second new pen pal, Brea, since the four students had started to get to know each other. She learned Brea and her mother liked to do botanical painting with watercolors, so she was looking forward to learning something new and creative. It would mean decontaminating after a visit, but she would probably be comfortable doing so after a little practice. Or she didn't have to visit their house. She had heard Cara baked, and she wondered how that could be if they didn't always have flour, so she would probably visit on a day they had electricity.

It would feel a little like home if she could do some baking.

Breathe, she had to breathe.

It would all be fine.

Wouldn't it?

HYPER-RICH
People Learn of
Rehousing Plan

Hyper-rich People Learn of the Government Plan 2028

"This was my greatest idea ever! Towers on high ground with walls around the acreage and drainage tunnels underneath will keep me safe from the coming climate change!" said the self-proclaimed richest man on Earth, Kingsley. He grinned at the day's news which quoted him as saying there is no such thing as climate change.

"You haven't heard?" asked a dejected Richey, as he held up a letter on his tablet. "I have court orders to share my living towers, like the ones you built, with the nearby population."

"Who cares about you?" laughed Kingsley. "I have no orders from the court. Forgive me a dastardly chuckle!"

"Have you been reading the new regulations? Have you been checking your mail?" asked Chase, choking on his words. "The government wants us to open our towers to anyone who wants to move there! For free," he whined as he straightened his long tie.

"The court plans to move the masses inside the walled cities we built to protect ourselves is outrageous! Those areas should be ours and ours alone!" complained Richey as he texted his on-call lawyer.

"Yeah, we paid to protect ourselves against climate change!" shouted Yates while his stock-broker waited on hold.

"But we are spreading propaganda that climate change doesn't exist!" [xxiv] moaned a stunned Richey. "And it is so convincing. It almost convinces me!"

"I just received a court order, too," protested Sterling as he pouted, stamped his foot, and turned red.

"You are right, I also have orders to let people live in my towers!" groaned Kingsley as he viewed his delivery digest from the online mail service.

"Does your court order say allow non-paying people to live in your apartments or you face trial?" exclaimed Sloan.

"Face trial? For being hyper-rich and taking care of ourselves? What's the crime in that?" asked Sterling as he rubbed his 24-carat gold company pen encrusted with diamonds between his manicured and massaged hands.

"Well, we did run our businesses in ways that increased climate change," mentioned Chase, as he rubbed the Botoxed area near his chin.

"Anyone would! Doesn't everyone want to be like us?" asked Belle while she twisted one of the 15 carat diamond rings on her fingers.

"Of course, they would! We are very rich, and we thought we could get away with it. How could poor people afford to sue us? They don't even know their rights!" demanded Sloan.

Yates proclaimed, "No one knocked on my door and announced fossil fuels were trapping more heat in the atmosphere. I thought everyone everywhere wanted a car that ran on gas. I supplied the cars and the gas! I was helping people!"

"Oh, really? Did you live under a rock?" asked Sloan as he looked over his stock portfolio "Greenhouse gases made a blanket around the Earth.[xxv] You thought that was good or something?"

"Well, you thought wind energy gave people cancer!"[xxvi] answered Yates.

"That was only propaganda I helped spread," admitted Sloan hitting sell on an unusually high stock. "I knew better."

"We all knew better!" muttered Richey.

"Whatever will we do? I can never live among the poor. I'd probably get dirty and sick. I can't live that way! Not ever!" sputtered Yates.

"I simply cannot subject myself to lice, darling. Never!" added Celestine as she applied her thousand dollars a tube lipstick.

"Say, Kingsley, don't you have a secret spaceship in case of something terrible happening? This is terrible," reminded Chase. "Remember our plan?"[xxvii]

"Yes, horrible, unthinkable, and I do remember now. I had a little 200-billion-dollar doomsday spaceship created a few years ago," thought Kingsley.

"Wasn't it the *Slip Away Spaceship*? Can we just slip away? Disappear into the night?" requested Richey.

"Yes, yes, it is high time. I'll have the spaceship dusted off. Prepare to fly in three days. Only this group and of course, the astronauts, and the workers we will need to attend to our every need will be allowed on board," answered Kingsley. "Activate the announcement and pack your designer space suits."

"How long will this trip be, again? How many years until we reach the uninhabited planet full of natural resources, all ours for the taking?" asked Chase.

"Thirteen years. It will go by in the blink of an eye with the entertainment system installed on board," stated Kingsley.

The hyper-rich wondered to themselves if their escape would happen before being arrested and how much money they could gather in a few short days.

Joe, the lead worker, spread the news, told the others to mobilize and start boarding the ship that very night. Everything must be just so for the billionaires to be comfortable as possible in outer space. Food, water, medical supplies, electronic devices, and games were ready. The fish aquarium was activated and stocked because, after all, the hyper-rich often required caviar. Astronauts and workers had signed contracts to work on the *Slip Away* several years ago, never thinking such a thing was possible. They had accepted and spent the money at the time. There was no going back now.

Not being allowed to say goodbye as per the terms of the agreements would be enforced by private bodyguards left behind. The crew and wait staff quietly slipped out of their lives to the secret location of the spaceship.

Bootsie and Lynch would do what was required to families of crew members who did not obey orders. About a year ago, Lynch had found some guy named Troy wandering around after a storm. After learning he was a technology specialist, he locked him in a basement to help prepare technology for the voyage of *Slip Away Spaceship*. Now it was time to force Troy to do the technical monitoring from Earth to outer space.

Each astronaut and worker thought of family and people they would leave behind and never know again. Perhaps leaving the planet was for the best, given the lack of oxygen Earth was facing. At one point, each had believed escaping Earth was the best idea. Now that it was actually happening, who knew?

There was no longer any choice.

It was a done deal.

PROTESTS &
CARA'S DECISION
2028

Decisions and Life in 2028

Cara was visiting her mother, Brea's grandmother. Brea was almost six years old and didn't understand why Grandma's house had a wet floor all the time. She was old enough to smell the mold and mildew.

"Come and live with us, Mum. Our house is closer to the high land walled city. It would be drier for you, and we would love to have you," suggested Cara.

"This house is all I have!" argued Grandma.

"Not really, look at the water seeping in as climate change causes rising water levels," said Cara, kindly. "It won't be standing for long."

"Oh, dear, it just is not drying out, is it?" complained Grandma.

Brea tripped in a puddle in the kitchen and fell. "Ouch! It is hard to walk in this house."

"Well, I can't stay here much longer, and I do have a suitcase packed," admitted Grandma.

"Then, it is settled. Let's leave now," Cara urgently said.

"How can I repay your kindness?" asked Grandma.

Cara told her about the protests at the courthouse regarding the hyper-rich and the upcoming trials. She suggested Grandma could babysit while she protested. The decision was completed, and Grandma took her suitcase and went home with with Cara and Brea.

Later, Cara went to a coffee shop and was checking messages about the planned protests and where to meet up beforehand. The roar of the crowd outside grew louder by the minute. She opened the door and stepped outside a minute to see what was going on. She was swept up into the crowd and pushed in the direction of the courthouse. People were holding signs and shouting, "Let us live! You ruined the planet; let us live!" The trials were scheduled to begin, the walled areas were still closed, and the hyper-rich had not appeared. The rumor was people were going to be allowed to live in the apartment towers rent-free, but that seemed unimaginable.

Shouts and cries and people looking up to the sky took over as people pointed to a spaceship taking off. Some people were shouting, "They escaped! They escaped?"

"Who escaped?" Cara asked a person standing by her.

"Who? The people too chicken to face court! It won't be long now! They will open the walled areas to as many as possible to live in the apartment towers!" he said.

"Where are they going?" she wondered.

"Who cares?" answered the man.

Cara ran home.

Arriving home breathlessly, Cara told Grandma to sit down. Expecting bad news, Grandma sat. "You won't believe what just happened!" Cara said.

"I heard a rocket or something, and I thought those were outlawed to help climate change," answered Grandma as she rubbed her hands.

"Just a minute, I will catch my breath and tell you what's going on, that was a rocket ship or something!" Cara said.

"The loud noise hurts!" complained Brea as she put her hands over her ears.

"It probably won't happen again anytime, soon, honey, don't worry," Cara said, patting Brea's black hair.

"OK! So, the people who have been polluting the planet and preparing climate change 'proof' living areas for themselves were supposed to start their trial," explained Cara. "They escaped planet Earth! They flew off into space! No one knows where they went. But now the government is going to decide what to do."

"Decide what to do about what?" questioned Grandma.

"About the climate change prepared walled areas left behind with tall earthquake-proof apartment towers on high land, and drainage under the areas to prevent rising water inside," Cara told her mother. "The government might let people live in the apartments, rent-free."

"What? I can't believe what I am hearing. Let's listen to the news. I want to understand," said Grandma as she turned on the emergency radio. They took turns hand-cranking the emergency radio to power up the radio.

The radio program was broadcasting the same information Cara had just shared.

"We will have to decide what to do," Cara said. "Apparently, we have a choice. Things are happening fast, and we should decide."

Grandma asked if they could move later after they had some time to think.

"Now or never!"

The two women looked at each other and knew they both loved living in a free-standing house more than anything. That is if it was not flooded and standing in water like Grandma's. They thought they had some time left at Cara's house.

They stayed, not realizing an opportunity to move would not come along every day.

Robert was six in 2028 and did not understand about walled areas and apartment towers. All he knew was his parents were missing. All he wanted was his parents and his Grandfather. He knew Grandfather was looking for his parents everywhere since Bomb Land Cyclone number 277 because their house was destroyed. After escaping the storm, Grandfather took Robert to his summer place, an earth-sheltered home. It was near a walled area at the edge of the forest.

When the choice came to move to an apartment or stay in the earth-sheltered home, Grandfather decided to stay, as he could talk to people who were traveling through and show a picture of the couple. He could tell people he was looking for family members. Maybe someone would know something. He felt moving to a walled area would cut him off from attempts to find Robert's parents, which included his only son.

They stayed.

Grandfather worried he had maybe given up his only chance to move to an apartment inside the walled city center. Robert didn't seem to remember escaping the hurricane, and Grandfather never brought it up.

The government could not prosecute the worst climate change offenders with no defendants. The *Slip Away Spaceship* which had let them escape was quickly forgotten, although videos and memes about it circulated for months. The government offered rent-free apartments to all who wanted to move to the city center walled areas. Not everyone wished to move. Each had his or her own reasons, such as not wanting to give up a home, wanting to find missing family members, or just being too independent.

The rules were clear. Once moved inside a center, it was almost impossible to get permission to leave. The contamination of invasive plants, species, and pollution outside the areas was considered too risky for city centers. For instance, poison ivy was growing at an alarming rate due to increased carbon dioxide in the atmosphere. Going back and forth between inside and outside the walled areas required decontamination procedures. The billionaires had maintained pristine areas for their golf courses, apartment towers, restaurants, and resorts. It was worth the effort to keep the areas as free of pollution as possible.

An extremely fast timeline was implemented by the planners, which was very unusual for the governments. Laws against vehicle travel by fossil fuels, gun ownership, and any new use of plastics quickly became law and were enforced. Governments around the world stopped fighting wars and took conflicts to court. War was just too costly to the environment. Numerous local laws were written to help prevent more pollution.

Carpenters, bricklayers, architects, engineers, and experts in solar and wind energy were assigned city center apartments first. These experts would help plan and execute standard living towers based on green energy.

Solar glass windows and walls were to be installed in existing structures. Wind energy would supplement electrical needs. Water would be recycled in each tower. Green and white roofs were added to existing structures. All roadway repairs and maintenance were stopped, and work and financial efforts were directed to sustainable living. The government had to be decisive and fast. It was. Buildings seemed to shoot up overnight. Citizens had never witnessed governments accomplish so much on such a jam-packed schedule.

People who were willing to move immediately would live rent-free. They would work to adapt the walled areas for more people.

Robert and Brea's families took their chances and stayed living outside the city center walls.

LAND BOMB
CYCLONE #277
2027

Bomb Land Cyclone 277, 2027

The holiday season had begun, and soon it would be time to ring in 2028. Robert's parents had been cooking all day. It was almost time to set the table and later open gifts. Gray clouds had hung in the sky all day, and the barometer was falling. Anxiously, they looked through the window.

"It is too cold for a tornado," hoped Robert's father, Troy.

"I think you are right," answered his mother, Kelly. "There should only be a little snow today!"

Recently, the basement had flooded a little. A few times, the power had gone out. Living in Moorhead by some of the 10,000 lakes which made Minnesota famous was a weather gamble. Some people had moved away after the 2009 Red River flooding.[xxviii] However, they spent their summers in an earth-sheltered home closer to the twin cities and felt they always had an alternative place to live.

"We couldn't have a hurricane here, could we?" asked Kelly.

"No, but there have been bomb land cyclones from Minnesota all the way to Colorado since 2019."

"Let's try to relax and wait for Robert and Grandfather to get home." Kelly nervously glanced around. "Could you call them right now?"

Troy called Grandfather's cell phone.

"Time to eat!" Troy said happily to Grandfather when he answered the cell phone.

"Looks like Robert and I are stranded in a blizzard or something," Grandfather shouted in the phone as the reception faded.

"Blizzard? Aren't you just a few blocks away at the toy store?" Troy shouted back to the phone. "Stay put, that is what they say to do, and I'll call for help. I'll call back in a few minutes."

Troy never called back.

Within a minute, a violent tornadic thunderstorm hit the house, causing part of the roof to collapse and debris to fall on the couple. The house slid down the hill it was on in a mudslide. The home was destroyed.

Robert's parents had known about weather-related events such as flooding, stronger storms, and drought from climate change. Category 4 and 5 hurricanes had increased in frequency,[xxix] although less often. Troy knew about flooding and land bomb cyclones.[xxx] But what could anyone do? Where were people supposed to live?

Grandfather and Robert were dug out of the blizzard's snow in a few hours. They went to the designated shelter for their neighborhood. It was very crowded, but no one there was able to communicate with or find Troy and Kelly. Grandfather called the police and reported them missing. He tried to get a room at one of the motels in the area, but they were all booked.

Near nightfall, Grandfather decided to drive to the summer residence, thinking it was the best place to take Robert and keep him safe. Troy and Kelly could easily guess where they were staying. He could distract Robert at first with the toys they had just purchased. Robert was only five years old, but even he knew something horrible had happened.

GRANDMA
2032

Grandma, 2032

Cara knew her mum wasn't well. The years of living in mold and mildew had affected her lungs. She had developed chronic coughing and obstructive lung disease. It had been a matter of time, and Grandma passed in the night.

Living outside the walled city made the natural burial less complicated. A sustainable alternative to other burial methods[xxxi] was what Grandma would have wanted. Neighbors which included a retired minister helped Cara and Brea with arrangements to bury Grandma near her own home. The house was now in ruin. There was enough land area for the process. She was placed in a biodegradable shroud and coffin.

They hoped she would rest in peace as she was remembered to be such a kind person.

A conservation burial which would have restored native habitat and saved endangered species could not be conducted as the area was muddy with flooding. They did what was possible, and friends and neighbors comforted them.

WALKING
PATHS
2029

Walking Paths, 2029

Looking out the window, Cara asked Brea, "What is going on now?"

"Rocks!" remembered Brea.

"Oh, I didn't think it would look anything like that, I thought they would light up," said Cara. "Oh, of course, they will at night after a few days when the rocks have absorbed solar rays. So, what will we use our electricity for tonight, the fan or some cooking? I'm so hungry, but let's eat cold leftovers, as the temperature is high and it is humid if that would be OK?"

"OK, Mum," Brea answered. "I like the breeze from the fan."

In the daytime, they could see the paths were beginning to take shape. The digging had been to install drainage pipes and tunnels so as the water rose the paths would remain dry. It looked anything but reassuring with piles of dirt and mud near the paths. Everyone would now walk and not ride to prevent the use of fossil fuels. It was becoming harder to breathe with less oxygen, and although people were adjusting, bike riding required too much lung power for many people. A hovercraft was rarely used.

Cruise ships, a thing of the past, were now permanently stationed apartments along the coast where the sea was encroaching on the land. They could be moved closer inland as the water level rose. Brea had never been on a cruise, but a ship that emitted as many emissions as a million cars in a single day[xxxii] was to be avoided by the environmentally conscientious.

VINTAGE
Climate Change Game
2019
2019

Vintage Climate Change 2019 Challenge

"What is this message?" complained Kalli.

"I barely know, but we have to send in our answers before each symposium. I didn't realize this part. Before leaving for American, school will take time while we are here," said Tristan, looking at the electronic message.

"You mean we have to do this before we go?"

"Yes, let's just get it over with."

"I'm packing!"

"Not yet, please, I'll help you with your question," said Tristan.

Question one read, "What is hotter, wetter, and smells like cilantro?"

He entered the question in the online search and found the answer.

Both Tristan and Kalli exclaimed, "Minnesota!"

"Who knew?" asked Kalli as Tristan sent in his answer: Minnesota.

"I think I saw the answer to my question on that web page too," said Kalli.

"Which is?"

"How will pine forests change in Minnesota? Look, here it is, further down on the page, it says the trees will be oak and maple in the future. All the ash trees will die." [xxxiii]

"Right! Better send it in!" suggested Tristan, so she did.

Both students independently checked later and sadly found those things had happened. Because of the emerald ash borer,[xxxiv] the ash trees did indeed die. They were startled to learn children in northern Minnesota didn't even remember pine trees looked like as they had never seen any. "Didn't people used to use that kind of tree for Christmas trees?" sighed Kalli, the tree hugger. "It almost makes me want to cry my lamps out."

MACE
Answers Vintage 2019
Question One

Mace Answers Vintage 2019 Question One

Mace opened his expensive electronic communicator and saw a new message about some irritating game. A game question? To answer before symposium one? Climate change was too annoying. First, people moved in on much of his floor in the living tower. His father had paid for all that space and owned it lock, stock, and barrel from getting a gigantic monetary award for helping to build some spaceship. That is until the government reassigned them to a fraction of their formerly own huge, beautiful space. How rude of the government! Who cared if the common folk would drown in their own homes? Who cared at all?

He forwarded the stupid question to his father, Lynch, and asked if he really had to do any homework.

"It doesn't matter how you answer," replied Lynch. "Everything is set for you to win on presentation day and end up with the most points of your class."

"Just checking," laughed Mace. "It pays to know people in high places," he said.

He sent in a silly answer, then turned his attention back to the more important electronic gaming equipment.

BREA &
ROBERT
First Day of Class Fall
2035

Robert and Brea Attend New Schools

"Class today, Grandfather! Finally!" Robert called as they woke up. "Let's go to the cafeteria for breakfast."

"Sure, the breakfast here in Tower 17, beats trying to cook," said Grandfather.

"Right, and they have better food than we could get outside the city center," said Robert as the two enjoyed the meal. "I didn't seem to hear you asking to have the canned food in our cupboard," he laughed. "The tomato soup will be good if we get sick or if the cafeteria is closed. I hope we won't need it, but it is good to have on hand."

"The cafeteria food is not like Grandmother's cooking, but better than we were doing! Have a great first day, and be sure to tell me all about it tonight."

"I will. Bye! I hope you remember how to text me."

"I might not, but we can practice again later. Bye!'

At school, which was downstairs, Robert signed in and was given a biodegradable name tag. He was wearing the school uniform, which was made of recyclable cloth. For his first Vintage Climate Change Challenge, he was asked to name one 2019 clothing brand that was eco-friendly. After some research, he had named Eileen Fisher. Although he had never heard of Fisher, she had admitted textiles were second only to oil in pollution at an awards ceremony,[xxxv] so he thought that was a good guess. The total points earned in the game would be revealed at the end of the semester.

He thought about Brea at her first day at City Center High 8763 and wondered how she was doing. Her challenge question was difficult with limited electricity and barely any battery power. When asked about reducing environmental impact for juice packaging in 2019, she wasn't sure of the difference between frozen, bottled, and boxed containers. She had never seen any packaged juice. What were juice boxes? She had no clue. She wrote, "I will start participating in this challenge after school begins. I do not have enough power to research the answer while still outside the city."

Robert thought Brea was honest, which might give her a half-point or something. Brea wore her head protection also covering her eyes, the outdoor boots, and the clothes she had been wearing for a few days to attend her first class at City Center High 8763. What did they expect, a fashion show? She left her boots in the outside community decontamination box, put her helmet on a shelf just inside the entryway, and went to the decontamination showers. She was provided with an eco-friendly student clothing, comb, soap, and towel. Every day would start like this. A bonus for her was that her own clothes would be washed while she was in class. Why wear clean clothes just to take them off? Each day after showers, she was also given a clean reusable water bottle and hankie. Once through the process, she signed in at the office and received some information along with her schedule.

She would be at lunch, English, and Science with the other students. And surprise! She was being allowed to work on her own ideas to help solve climate change and was being offered the use of a research secretary to help her. Her? Oh, the school apparently knew what she, Robert, Tristan, and Kalli were learning on their own. The secret project was no longer confidential. At least, this meant their project ideas were being academically supported. The schedule said to call the independent work "study-hall" in the event anyone inquired. The other students would think she was being tutored or was behind in some subject. What they didn't know wouldn't hurt them, but she would have to be careful not to discuss the projects with anyone other than teachers and the four pen-pals, soon to be real-life friends.

Brea found the study room. It was one of a dozen small rooms off a narrow hallway. It looked fairly bare through the window but was very advanced technically when she entered. She left, looked again through the window to a plain room, and found the same technology when she went inside. The window was actually a hologram screen. Her research secretary was a robot. There were two chairs, two computers, one round table, and a 3D printer. That first day, she couldn't imagine how the printer and robot could be useful.

She couldn't wait to talk to Robert, but the first symposium was not for a whole month. She would use their secret language to communicate. It wasn't exactly pig-Latin but confused adults. How would she let him and the pen pals know they were being monitored? They had no plan for this eventuality.

She decided to send a coded message to the three others to avoid having to wait to tell them. She wrote the following.

Pupatuttutenotutionotutotuthatisusmooesussusa gugetuthatetuteacookhaterursuskooknoowowou rursusecookruretutpuplulanodudnootutwoworur ruryuk

She received three replies within a few minutes. Tristan's reply read:

Wowowowwowhatokooknoewow

Kalli's response read:

Mooyukdudadudtuthatougughatsuso

And finally, Robert's answer was:

Susususpupecooktutedudasusmooucookhat

Relieved they understood, Brea turn on the school computer to begin to study. She found several articles and a page of questions already in the documents. She began reading and researching for answers.

TRISTAN & KALLI
Travel to City Center 8763

Kalli and Tristan Travel to City Center 8763

Finally, it was time to pack and leave!

Kalli and Tristan packed few clothes as they would be wearing the eco student uniforms when at school in America. It was suggested they wear them all the time they were awake because they would be inside the city center walls. It was fine with them as it made for light packing. Kalli was hugging her suitcase which had a birthday gift from her parents as she was turning twelve on the day of the first symposium. All she knew about her present was that it would pass decontamination inspection. On the way to the airport, she was already missing home. She wondered why she was going. Oh, well, too late now. It had sounded so good at the time.

Tristan looked at her and just knew she was worried. He was too, but since he was about a year older and already twelve, he wanted to set a good example. "It will be fine once we get there. We could never otherwise ride on an airplane, you know, eh?"

Sometimes their mother's habit of adding an "eh" at the end of a sentence showing her Canadian roots. Tristan and Kalli had picked up the same habit.

"Yeah, I know," Kalli said softly.

They settled in on the plane and were cared for by the attendant. The food was different than they were used to, but it was good. Because the plane left about midnight, they slept off and on during the flight.

The hovercraft ride from the airport to the city center was much noisier than they expected. They gladly wore noise-canceling headphones. It took a few minutes to reach their temporary home, Living Tower 17. Decontamination wasn't too terrible, just a shower and change of clothing. They were each given a protective half-helmet to wear if they left the tower, while, of course remaining inside the city walls.

Kalli still had hopes of visiting Brea's house and wondered how she might get the protective boots she would need.

The other students would know they were the kids from Northern Ireland because their images had been shared electronically. That might help them meet people. They were given keys to their apartment and went straight to bed with jet lag.

Their classes would begin in two days, to allow them to adjust to the different time zone. They would be starting school during the second week of classes. It would be difficult to be new students that day, but also closer to symposium one when they would see Brea and Robert.

FIRST
SYMPOSIUM
Robert, Brea, Tristan & Kalli
Meet at School

Robert, Brea, Tristan, and Kalli Meet at the First Symposium, Fall 2035

The day of the first symposium had finally arrived, and today the four friends would meet. Name tags on, students found their assigned seat at a circular table in a large room. Arriving early as he lived in the building, Robert easily found his spot and waited. He watched as hundreds of students entered the room and puzzled over finding their correct seats. Finally, Brea arrived a little late from decontaminating and then walking from her own school just within the walled area. The two friends smiled while they chatted, poured water into the four glasses provided, and silently watched for the two students from Northern Ireland.

"I recognize Tristan from his picture!" Brea whispered as she stood and waved her hand high in the air. She finally caught Tristan's eye, and he walked over.

"Bout ye!" he said.

"Bout ye," answered Brea and Robert as they laughed at the inside language knowledge.

Right behind him about as close as his shadow walked Kalli. Smiles and handshakes all around were quick as the event leader had begun speaking in the microphone. Brea and Robert whispered happy birthday to Kalli.

"We are lucky our table is only for four people," whispered Tristan, smiling again. "We have a project to keep secret."

"Shhh!" whispered Kalli. "Don't be an eejit." The word "eejit" needed no translation.

The leader of the event was one of the teachers from Brea's school, Mr. Washington. "We have a special day today and hope all students will make the most of the time available. The agenda is in the center of each table. Please look it over as you meet any students you might not know at your tables. The first lecture will begin in ten minutes."

"Well, ten minutes is better than nothing. I just want to say it is fantastic to finally meet you in person, Tristan and Kalli," said Brea.

"Grand!" said Tristan.

"Brilliant!" added Kalli.

Everyone relaxed due to the little break in the first part of the day.

Robert asked if everyone had been working on the game they had thought of and would lead for the last afternoon session.

"You betcha!" laughed Brea using her best Minnesota accent, and laughing. "Dontcha know?"

A brown envelope with their game was in Robert's folder. It would be played like Jeopardy. Besides being some fun for the day, it offered them extra texting time when preparing game questions during the past few weeks. They quickly reviewed the game rules.

Looking at the other three teens, Brea was glad for the secret code message plan, and that they understood their secret plan was known to the teachers.

Across the room, a pair of green eyes were staring at their table. Mace was chuckling to himself. Not wanting to be too obvious, he pretended to read. The one rule he had was not to tell anybody about the plan for him to win. That wasn't too difficult as he had no friends, and could discuss it with his father.

The day went on with lectures and presentations. Speakers included those who had become famous when they were just teenagers. Clare O'Beara, a tree surgeon from Ireland, also gave a speech. It was inspiring to hear live presentations.

When it was time for the large group game, it went well and helped everyone finish the day with a smile. It was called "Climchardy" as it was a mix of letters in Jeopardy and changing climate. The room divided in two parts by moving chairs and tables, to form an aisle along the center. The teams named themselves the Sparks and the Gems.

"Bout ye!" the four student game leaders said together.

Brea explained the phrase was Norn Iron for greetings. Kalli explained the game rules.

After a coin toss, the Sparks were first. Robert asked the first student to name a category and amount.

"200 for fossil fuels."

"Freedom molecules," Robert stated.

"What was the fossil fuel natural gas called in 2019 by some people in Washington D. C. including the president?"[xxxvi] Cody answered in the form of a question.

"Correct. Student 2, which category do you select?"

Brea could hear one student whispering loudly, wondering what was up with all the 2019 facts.

"Ah," said Tristan as he had overhead the student comment and decided to address his answer to the entire group. "2019 seems to have been a turning point year for the worse in climate change, of course. The tipping point! it was the year of the Amazon rainforest fires. The pandemic began then, too. It was a time of change."

The game continued, and the other teens realized they were learning helpful information for upcoming questions in the Vintage Climate Change 2019 Challenge, so they paid close attention. Listening now would help reduce researching time for the questions they would receive before each symposium.

Everyone (except Mace who had already left the symposium to play more online games) sang "Happy Birthday" Kalli before being dismissed, bringing a few tears to her smiling face.

KALLI'S BIRTHDAY

Away from Home

Kalli's Birthday Gift

After receiving a pass to go upstairs and see where Kalli and Tristan were living, Brea went upstairs to visit guided by Robert. He knew his way around the building and needed no pass because he lived in that tower, too. Kalli was waiting to open the birthday gift from her parents until her friends arrived. There were some vegan cupcakes on the counter in their small apartment that some kind soul in the cafeteria must have left there, knowing about the birthday. The cupcakes would help the celebration.

The apartment had plain furniture and décor, an immense window, and top-notch kitchen and bathroom fixtures. Built for the hyper-rich, the woodwork, flooring, even the ceiling was beautiful. The students would have no complaints!

Brea gave Kalli ten pieces of watercolor paper, some of her paints in small glass containers, and a paintbrush. "I wanted to give you ingredients so you could cook, but you will visit my house soon," Brea said as she wished her a happy birthday.

"How did you get these art supplies through the inspection?"

"Actually, I sent them ahead my first week of school and picked up the package today. The administration already took care of any decontamination necessary," said Brea.

"Thank you for going to such lengths for me!"

Tristan and Robert gave Kalli some solar charging batteries, laughing suspiciously as they did. "What's going on?" she asked. The laughing helped keep the tears away as she thought of her parents at home in Northern Ireland.

Kalli picked up the gift from her parents and unwrapped it, having no idea how it had passed inspection when she saw it was a drone. A drone? The school and city administrations had obviously approved the device before she even arrived.

"Maybe they thought it was a scale," suggested Brea. "I've seen scales that sort of look like a drone in-between clear plastic squares."

"I bet they knew, even if we didn't," added Robert.

"A drone?" Kalli exclaimed. "Now I really don't know what is going on, but it is a nice gift."

Surprised, Tristan answered that made four of them. "But I have a feeling we will soon find out!"

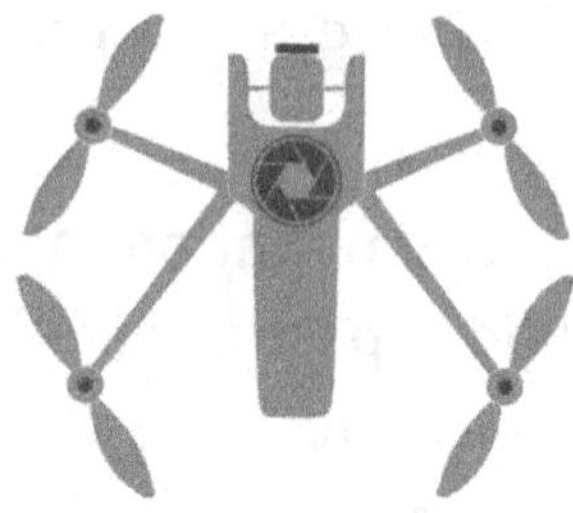

The cupcakes were a special treat and nice to share before they had to call it a day. The four students wouldn't be able to gather together again for an entire month. Apartment visits were limited to symposium days.

SOUNDS
In the Woods

Brea and Kalli Notice Sounds in the Woods, Late Fall 2035

Finally! Kalli was going to visit Brea's house a little way outside the city center walls. She found the fourteen feet high walls a little threatening when she was close to them. She was proud to have a pair of boots for walking through the invasive species while wearing her half-helmet. She happily walked along as she realized she was walking more slowly than she had been inside the city area. She was breathing a little harder and getting a slight headache, too, due to the depleted oxygen. She wondered if it had been that way at home. She hadn't noticed before. Thrilled to see her friend's home and meet Cara, she followed Brea's lead.

The two girls put their boots in the decontamination box, placed their head protection on the shelf inside the door, and went into the kitchen. Kalli looked around at the wooden shelves filled with plant pots which held attempts at growing food indoors. She especially noticed carrots and green beans. She saw peppers hanging from a string nailed to the ceiling, and dishes in the sink. Even she knew water was saved to wash as many dishes as possible at one time. Clean water could not be wasted as it came from rain, not the ground anymore.

She was happy when she saw the oven!

Not being able to help herself, she asked if she might be able to help cook dinner. She was itching to work in a kitchen.

Cara assured her she could, after the watercolor session. Cara had selected safe plants for them to trace before painting. She taught them how to "stipple" to add texture to the flowers and leaves. She showed them how to shade petals, so they looked rounded and almost three-dimensional on the flat paper. Kalli had never tried realistic painting before and was fascinated. She realized that even in the woods away from the city, there were wonderful things to learn.

Kalli asked, "How did you learn to grow so many plants for food indoors?"

Cara answered she was friends with the florist at Whole Foods back before 2020. Sarah was the botanist and florist at Whole Foods Market. When Cara asked questions, Sarah kindly answered them. "It was most helpful when I had to start growing our own food indoors," explained Cara.

"Do you know where Sarah is now?"

"No."

"You look sad!"

"Yes, but maybe one day we won't have to live with so many restrictions, and I can contact her again."

"Well, we don't have much time, let's get started!" Kalli showed Cara how she baked cookies with chickpea flour, as an hour of electricity for using the oven had been saved for this activity. Then, Cara's soup made from food grown indoors was shared. Everyone agreed the baking and dinner were delicious. After dinner, it was time for Kalli to return to the living tower.

Brea starting walking Kalli back to the guarded door in the wall.

There was a sound in the woods, causing the girls to glance around. They weren't alone. Brea whispered, "Did you hear that?"

"Yes, and I saw a shadow behind a tree way over there," Kalli made a slight motion to the left towards the forest store. They waited and were silent and decided it was no big deal. They continued walking.

Kalli placed her borrowed boots in the designated container and hugged her friend goodbye. Then it was decontamination time. Everything had been so wonderful it didn't bother her very much to go through the process.

Brea hurried home as fast as she could, in case the noise had not just been some animal.

SECRET
PROJECT
Will it Stay Secret?

Will the Secret Project Stay Secret Until Presentation Day?

Soon the Vintage 2019 Climate Change questions arrived prior to symposium two. Time was flying by as the students researched and made decisions about their individual and group projects. The project proposal presentations were due on the day of the second joint symposium. Funding and support from the school, or even the city administration, could make or break a project.

Tristan laughed at his question, as it was about peat bogs in the UK.[xxxvii] He had been handed one easy to answer. He wondered if that was on purpose to save him time?

> *Which peat bogs are causing climate problems like cars and factories, and how might this be solved?*

Tristan knew draining peat bogs released CO_2 and it could be solved by blocking drainage ditches and restoring vegetation. Although depending on where the bog was located, the costs could be high, the benefits to the Earth would be invaluable. He sent in his answer without even checking as he had already studied this issue as a possible project. He had already decided his individual project would not be about bogs, however.

He wanted to draw a climate credit card and write a humous answer. But he didn't know the exact costs and the answer had to be text only. He probably should not try to tell an answer as a joke.

A little while later, he received a strange message which said peat bogs cover 12% of the UK. He couldn't tell who had sent it.

Hmmm? Was someone reading his messages?

Kalli's question was also easy.

> *Why should people take walks in the morning instead of evening?*

She knew there was more oxygen in the morning from researching poison ivy and other plants before she visited Brea's house. She answered that plants give off less carbon dioxide at night, so the morning air contains more oxygen.

A little while later, she rushed to find Tristan and tell him someone she didn't know had messaged her asking if she wanted to take a walk in the morning. The two decided they were being watched electronically, and not by the school or city. Someone was letting them know their files were not safe. Just when they had reached their final decisions on project proposals and had them ready to turn in to Mr. Washington!

Kalli wrote to Robert and Brea:

yukoububecookarurefufulul

Three replies all said

Ohhatnooh

A little later, four electronic messaging devices received the same message:

Hatahata

Their devices had obviously been hacked. Who would do such a thing?

On board the Slip-Away, Kingsley laughed when Joe accessed the computer messaging system and told him the bodyguard's tech help on Earth had completed the latest task. The students' secret code was cracked, emails were hacked, files stolen, and other nefarious tasks were complete.

Joe felt sick inside and decided not to transmit any more messages from Kingsley to Earth. He would not be admitting it to anyone on the ship, though.

Spying on kids was just wrong! He had young relatives on Earth and didn't want them to suffer. He hoped the bodyguards didn't have another way to communicate with Kingsley.

He wondered if he could trust someone named Troy who sometimes sent messages? He hoped so.

MISSING
FILES!
A Mystery

Missing Files!

At the second joint secondary school symposium, the four friends were working in a computer lab reserved just for their team. Logging on revealed their files were missing. All that research, gone! All their struggles over project decisions, gone! Their computers and the server had no documents left. Had they done something wrong? What was going on?

It was time to talk to Mr. Washington about the missing files and also the strange electronic messages showing they were being watched.

Mr. Washington said he would find out what he could and see if any backups could be restored. The students were told to wait in Tristan and Brea's apartment until further notice. He wrote a pass for Brea. The four friends went upstairs and kept the door locked while they waited.

They were supposed to present their final project proposals during this symposium. They were ready to share their information and plans with the other students. The minutes passed slowly as they waited for Mr. Washington's answers.

STRANGE QUESTIONS

For Robert & Grandfather

COMPOSE

Inbox (6)

Done

Drafts

Important

Sent

Spam

Trash

Strange Questions for Robert and Grandfather

Grandfather could not understand why Robert had not texted back to him when he had written about the strange messages he had received. Wasn't the point of the device so they could keep in touch when Robert was in school?

The messages had questions that upset Grandfather. They asked the year of his birth and Robert's middle name. Grandfather knew not to divulge such information online, but the messages were signed by a Troy. Grandfather tried not to get his hopes up that this Troy could be his missing son.

Where was Robert? Why wasn't he answering?

Oh, he remembered, Robert was presenting his individual research and plan today at school. That was probably what was keeping him away from his phone. Maybe he should go downstairs and listen to the presentation. He went downstairs, found the large room, and took a seat. He noticed a student with green eyes who seemed to be laughing at him. *Weird kid, thought Grandfather.*

Looking around the room, he didn't see Robert, Brea, Tristan, or Kalli. Weren't they at school today? Worried, he went back upstairs to see if they were in Tristan and Kalli's place. But why would they be there on this special day? Were they playing hooky or something? It didn't seem they were the kind of students to do that, though. No, he felt something was wrong.

Upstairs, he knocked on Tristan and Kalli's door. He heard some whispering, so he knew someone was inside. "It's Grandfather!" he called.

The door opened a tiny crack and then more so he could enter. Kalli locked the door behind him.

"What are you doing up here? Shouldn't you be in school?" demanded Grandfather.

Robert put his finger to his lips and whispered an answer, which of course, the older man couldn't hear. Robert wrote on a piece of paper to communicate.

"What! Why on earth would anyone do that!" yelled Grandfather.

The students each put a finger to their lips. They all sat down and waited.

Robert also wrote a message to his grandfather that they didn't go to his apartment because he did not hear well, and if they raised their voices, they might be overheard from the hallway.

"I'm so sorry," whispered Grandfather, as he looked at the sad faces. Then he remembered his strange message and wrote down what had happened to him. It didn't seem like the two kinds of messages were related, but Robert wrote down that Grandfather should show the messages to Mr. Washington.

PRESENTATIONS
And a Blackout!

Individual Presentations Symposium Two and the Blackout

Individual presentations for the students from City Center 8763's Secondary School and the Living Tower 17 school had begun without a hitch. Students presented their work, much of which was pretty dry. Many students thought more trees could be planted. Less fossil fuel had been used for many years already. Some students suggested more drainage ideas. The usual kinds of ideas.

The problem with planting more trees along coastal areas near the sea was the rising water had a higher concentration of salt, which affected plants and trees. Some students only realized how saltwater might be a problem after presenting their ideas to plant along the coast during the time allotted for questions. It didn't occur to them when speaking.

One student, Mace, was very happy when he noticed some of the competitors for monetary awards were missing. He had already changed the name on a few papers and combined them into one document. His father, Lynch, would be so proud. Smiling with how easy that was, he would be presenting fairly soon. Everyone else was bored and wondering how a day could be so long.

Suddenly the electricity went out. How did that happen with solar and wind power? School was canceled for the rest of the day, but students were to wait on the lawn for an announcement to go home.

Mace stamped out of the school, but couldn't divulge why he was mad to anyone. And he had been told along with everyone else not to return home yet.

Up in the apartment, Grandfather and the four friends were suddenly sitting in the dark. They walked over to the window. It was a sunny day, and they could see students were out on the lawn. They seemed to be waiting. Maybe the entire building was in the dark?

They froze in place when they heard a knock on the door. Mr. Washington called in softly to let them know it was him. Opening the door, they could see by the look on his face that Mr. Washington had something to say.

"I know what is going on, and we are working on fixing the problem."

"Grandfather has something to tell you, Mr. Washington," stated Robert. "He received a strange message."

"First, let me explain what we figured out about your missing files, is that OK, Grandfather?"

Grandfather nodded yes.

"OK, this is a rather long story, but I'll make it short for now. The school administration cut the electricity to stop electronic transfers while files are repaired and things are back to normal. Do you remember when the *Slip Away Spaceship* took off, around 2028?"

Grandfather nodded yes and added that the teens were probably too young to remember. "The hyper-rich left the planet, right? How does that figure in today's problems?"

"Kingsley, the mastermind of the *Slip Away Spaceship*, had every minute detail planned for the escape and back on Earth for years and years. He took dozens of crew members with him such as astronauts and servants. He even planned for any people who said they would and did not get on the ship to be killed or captured by his private bodyguards left on Earth. Everyone had been handsomely paid, and most of them spent the money immediately, so they had no choice but to obey. They had signed contracts," Mr. Washington explained. "We know who the bodyguards are and have been watching as there have been some unexplained events. We suspect it was one or more of them who hacked your devices and files."

"Why us?" asked Tristan. "And why would they want to stop improvements in Earth's atmosphere for some people up in space?"

"Well, I know you know the teachers and city officials know what you are working on, and the bodyguards are supposed to ruin brilliant plans such as yours. Kingsley wants the Earth uninhabitable. Apparently, the bodyguards had some sort of contract with the people up in space. We know that but not why they would be doing these things now. Of course, we suspect they were paid handsomely."

"He did a good job of that," Grandfather said sarcastically, "Polluting the planet and then leaving. He was a terrible man."

"Right and even having people sabotage climate change efforts the past several years," said Mr. Washington. "Well, there has been a breakthrough. Supposedly, the hyper-rich do not know how to access the communication equipment onboard the *Slip Away Spaceship,* and they have been relying on Joe to read messages from Earth to them. Joe has gone rouge. He has passwords the people on the ship do not know, and he has stopped telling Kingsley and the others the truth. But he can't help us much, he is stuck on board with them another six years and after that on the new planet."

"This means what, exactly?" asked Brea.

"We know it was a bodyguard who stole your research, project plans, and presentations . . . and gave it to his son."

"Don't tell me, let me guess, could the son be someone named Mace?" said Brea.

"Right, but we haven't caught him, so we want you to stay in this apartment or Grandfather's until we find him," added Mr. Washington.

The lights went back on to everyone's relief.

"Are we in danger?" asked Grandfather.

"Not if we can catch Lynch and Mace first, so just sit tight. We will notify you when we get things back in order. Because the lights went on, I know their computers and files have been confiscated by the peace guards, and are now in safe hands."

Robert mentioned Grandfather's strange messages, which Mr. Washington was surprised and glad to know about and said those would be reviewed next.

MEANWHILE ON THE *SLIP AWAY* SPACESHIP

"Everything fine, Joe?" asked Kingsley aboard the *Slip Away*. Mace presented the best plans and won the financial award, right?"

"Right as rain, Boss," laughed Joe.

"Forgive me an evil chuckle!"

"Sure, Boss, I always do."

Joe thought he would also forgive himself for not telling Kingsley about the sunken bank vaults or his change of heart about helping the people of Earth.

SYMPOSIUM 3

Revealing the Secret Project

Revealing the Secret Project Symposium Three

Symposium three began well, with one missing student, Mace. However, Robert, Brea, Tristan, and Kalli's computers and devices had been restored and the files found. They had only two weeks to finish preparing as it took almost two weeks to get their original work back together. Their files were safely being monitored at all times, as well.

Lynch and a few of the other bodyguards had been found with stolen items and electronic devices and spy equipment. The other bodyguards left behind had been questioned. They were happy to find they could get out of their contracts with Kingsley, which were now null and void due to Joe's change of heart. Everyone but Lynch willingly turned in their devices. Clever technicians would be sending communications to the *Slip Away* to disguise what had happened. Joe would just have to figure it out.

The four students were ready not only with their individual presentations as prepared for symposium two but also with the group presentation and plans. Fired up and ready to go!

Grandfather was in the audience, and the group first presented individually. He was happy to listen, although his strange messages had not been solved. Maybe it was some sort of spam, but that name "Troy" had Grandfather thinking about it a few times a day. Oh, well. It was time to listen to Robert's presentation.

Kalli and Tristan wanted everyone to know that although millions of trees had been planted since 2028, the resulting forests had to also be responsibly managed. Trees alone could not solve everything. Some trees had been planted too close together, and some had been planted in now waterlogged areas. Their plan suggested forest maintenance.[xxxviii] For instance, in 2019, a law in the Philippines required every student plant ten trees in order to graduate,[xxxix] but the trees also needed ongoing care.

Sustainable forests contributed to the health of the trees and gave them enough room to grow. They showed a slide comparing two trees that were 75 years old, and the one without enough space to grow had a thin trunk, spindly branches, and few leaves. The other tree with enough space to grow in had a thick trunk, strong branches, and healthy leaves. This slide made a lasting impact on the audience, as many students didn't realize cutting down *some* trees could actually help the planet.[xl] Some of the slides were created with help from images captured by the drone. Kalli posted their free presentation online and shared the link on the Students of the World Unite (SWU) web page. Some large school districts had asked for permission to post the material, as well, which was an honor.

Brea had become interested in bogs after hearing Tristan and Kalli lived near one in Northern Ireland. After reading and finding out that when governments officially named bogs meant no one could remove and use the peat, a fossil fuel. Drainage ditches could not then be placed in a bog. She quoted what Tristan taught her, that bogs have twice as much organic soil carbon as standing forests. She didn't think she had ever seen a bog before, and by using Kalli's drone, she was able to get an idea of what it might look like in real life. She flew the drone until it was almost out of range to see the Sax Zim Bog near northern Minnesota. Her presentation was free online, and the link shared on the Students of the World Unite (SWU) web page. A few city administrators were very interested in her work.

One question asked was, "Are there dead bodies in the bogs in Northern Ireland?"

"Yes, and in other countries, too. That isn't the point of my research," Brea patiently answered. "It is an interesting topic. I included a link in my online article to information about bog bodies in Denmark you might like to read." [xli]

Robert had wanted to work on something special because of the problem his parents had faced when the house slid down the hill and was destroyed. He still missed them very much. His invention helped people know when a landslide might be on the way, and they should leave their house to be safe. He had borrowed Kalli's drone and souped it up with a little laser device for some of his work.

> "Lidar is a laser-based technology that allows a geologist to not only precisely and accurately locate landslides but also reveal its history and give clues to its makeup."

He had invented a new generation of lidar technology and demonstrated it outdoors with the help of the geology teacher. [xlii] The students were amazed. They saw printouts of what the land they were standing on looked like previously.

One of the questions asked after the four presentations was why they had taken an extra month to do their work. "Won't they be punished for turning in their work late?" asked a student. The teachers said no, they wouldn't, and they had missed the deadline due to technical difficulties beyond their control.

Sadly, Mace wasn't able to participate in the symposium today as he was busy doing work as the only student in an all-day individual study hall. He had been taking this special "class" for about a month.

THE SECRET PROJECT

Of course, the most exciting presentation of the day was the one the four students had worked on together. The other students realized before they finished presenting, they had not been being tutored for being too young or too far behind academically for secondary school. They realized something wonderful had been going on while they were working on shorter projects, and that they probably would not be winning the funding everyone hoped to receive. Their faces went from realization to intense concentration as the secret project was revealed.

"Nanoengineering makes many new things possible." [xliii] began Tristan. Right there, the other students realized he was far ahead in his STEM education. "There are still big environmental problems in the world," he explained, "but some of the solutions are tiny. Nanomaterials have been used to remove carbon dioxide from the air and capture toxic pollutants from water."

"We were looking at another way to use them," said Robert. "Different."

Brea added, "My mother and I live outside the city walls, and one of our volunteer jobs is pulling up poison ivy plants. We never knew what happened to the plants, but all we knew was they could not be burned or eaten. As more and more carbon dioxide spread in the atmosphere, poison ivy thrived."

Part of the video was shown. This part showed Brea and her mother pulling up poison ivy plants, including the roots, and putting them in collection containers. It showed them going to the outdoor decontamination shower, hanging their work cover-ups on tree branches, and leaving the containers with poison ivy nearby.

"We now know what happened to the poison ivy that was gathered," said Kalli, holding up her drone camera. A few of the students gasped with anticipation. "We learned what happened was . . . "

"NOTHING!" all four presenters said at that cue. "Nothing."

The students in the audience leaned forward, intently listening.

"By using the drone, we located where the ivy was being held, which was in former swimming pools.

It was just sitting there, causing allergies," said Robert. He pointed at himself when he said allergies, which got a few laughs.

"And we tried something, using the drone. We tried nanoengineering," said Tristan as he held up a small black spongy substance. While holding it in a gloved hand, he lit it on fire. It burned, and the remaining substance stayed about the same size.

At this point, he played another part of the video taken by the drone after obviously having dropped more of these carbon black sponges, all on fire, on the collected plants. "Do you know what happened?"

"Urushiol is oily, right?" asked Michael. "So you were able to absorb it, thereby reducing allergens?"

"Right!" said Tristan.

"Why didn't the sponge things absorb the water? It looks like the plants had plenty of dew on them," asked Alicia. "Why did they keep burning?"

"These nano carbon sponges hate water! They love oil!" Tristan explained. "Just look at the plant remains after about an hour."

"But we have, like, millions of poison ivy plants all over the planet," complained Ezra.

"Right, and we can remove much of the urushiol by using the sponges."

Another minute of video was shown.

"How did you get those sponges?" wondered Mr. Rodrigues.

"I used my 3D graphic design abilities and a 3D printer to make them," said Tristan. He didn't mention he had been able to visit Brea's study room with the 3D printer one day, and the robot secretary had been put to work after that cranking out hundreds of the nanoengineered sponges day after day.

"What are the rocks on the top of the plant remains in part of the video?" Trevon wanted to know.

Brea took that question. "Those are some of the rocks my mother and I collected in the mud, off the solar paths. We would pull plants and often find loose rocks. We noticed poison ivy did not grow close to the solar rock paths. That made us wonder. We think the rocks interacted with the poison ivy and helped them decompose without releasing too many allergens after the oil is absorbed. We don't know why, but it seems to work."

Tristan added, "We did research, and it appears crushed basalt could have cooling impacts on the environment. The rocks used on the paths are basalt. We hope to learn more about the process in further studies."[xliv]

"But! How will you get so many sponges like that?"

"They are reusable."

The four students received a standing ovation.

PROJECTS
BEGIN
Seeing Results for Earth

Bringing the Projects to Life and Seeing Results

Projects were uploaded, shared for free, and were watched world-wide. People realized more things they could do for the sake of the environment and began doing what they could to help. Using less fossil fuels, no more war, and planting trees had all helped. But people hoped to be able to leave the city centers and at least ride bikes in the future. Families had been separated for years in many cases and wanted to reconnect.

Scientists began working on nanoengineering and poison ivy, closely following the students and reviewing their findings to replicate their success in other places. The students even contributed part of the money they won to help fund the project. Everyone was hoping for a more livable Earth.

The remaining joint symposiums four and five were dedicated to implementing the final winning projects. Some students wrote letters, some sent electronic messages, some answered questions on the online presentations, and some were busy with 3D printing as Brea's printer had been moved to the school in Tower 17. Everyone was busy helping in whatever way they could.

Clare O'Beara helped a group of scientists and students work on the problem of how saltwater affects tree and plant growth along the coasts. They began sharing and working on her solutions.

Slowly, the immediate areas outside the city walls became drier. It seemed less poison ivy grew. And the temperature of the planet went down a half a degree in the space of a few months. Things were turning around, not just remaining in a holding pattern as before.

STRANGER

Knocks on the Door

Stranger in the Woods Knocks on Brea's Door

There was an unexpected knock at Brea's door. She and her mother did not have much unannounced company, and they were surprised. A man was standing outside. He was not wearing decontamination boots. He seemed desperate. What should they do? Brea just had a feeling this person might be related to the strange message Robert's grandfather had received. The message was not related to the sabotage of their projects and did not seem to be spam. To Cara's horror, Brea began to speak.

"Shh, Brea."

Brea took a chance and called through the door, "Troy?"

"Yes," came the answer.

Being careful, they took him to the outdoor decontamination shower and helped him enter the house without his shoes. Even though he was wet, they welcomed him in and gave him tea.

"Who are you?" asked Cara.

"I'm Troy, Robert's father," he answered.

"How did you find us?"

"I saw my picture on the bulletin board at the store not too far away. What was that? The One-Stop-Shop? It was at the edge of the forest. I asked who put up the photo of Kelly and me. The store owner made me take off my shoes and socks to talk. He said Brea was a friend of Robert's. Then I shopped for a little food."

"And to think stores used to have signs saying 'No shoes, no service.' Where have you been?" wondered Cara.

"Well, I don't know if you can believe this, but after our house was destroyed in a mudslide, I was in shock. A neighbor said he would help me to get to a hospital, but I was taken to a building where I was forced to live in a locked basement for a few years."

"Since about 2027?" asked Brea.

"Right."

"Why didn't you contact Robert or your father or the peace guards?" Cara asked.

"Well, I was a prisoner of sorts. My technical expertise was needed by some people who were going to leave the Earth on some sort of spaceship. They wanted me to hack all sorts of bank and social accounts. I spent my time guarding their funds and properties online, although I think they will never be able to return to Earth."

"Ah! Were you one of the Earth bodyguards?" Brea asked suspiciously, sitting up very straight in her chair.

"No, nothing like that, but I do know about that. I was kept at the building and given a basement apartment there, but I could not leave or communicate with anyone. I was able to get a little help from someone named Joe on the spaceship, but we had to limit our messages."

"What happened to Robert's mother?"

"I was told she was dead by my captives. Have you heard from her?"

"No."

"Robert must be so sad. Well, recently, something happened, someone had a change of heart, and I was able to escape. One of the Earth bodyguards for the spaceship practically opened the door for me, a guy named Bootsie. Still, I was given no information or help with how to cope with life now. I'm just figuring out walled communities don't just let anyone inside. By anyone, I mean me."

"I have an idea! Listen up, both of you," said Cara. "And then we will catch you up on what has been going on with city centers and resisters."

ABOARD THE SLIP AWAY SPACESHIP
They are Sorry Now!

Aboard the Slip Away Spaceship

"This is my greatest idea ever! I'm so awesome! Escaping to a new planet! Here we are seven years away from that polluted place, safe in outer space on the *Slip Away Spaceship* and on our way to a new, uninhabited planet. We will be able to use all the natural resources we want to our heart's content! Here's to me!" proclaimed the self-proclaimed richest man from Earth, Kingsley. He spoke as he checked his hair in a mirror to be sure it looked perfect.

"Yes, Boss, great, great!" affirmed the workers' leader, Joe. "It will be a wonderful new life!"

"Sneaking off the planet was brilliant. I've got to hand it to you, Kingsley. We escaped the worst of the trials," said Slone.

"Slipped off into the night, didn't we?" said Kingsley, pleased with himself.

"We aren't suffering through climate change anymore, along with all those regulations," said Chase, crossing his ankles and admiring his expensive shoes. "Now we can mine and drill all we want. Rich, rich, rich! We will be so rich!"

"Why the courts thought the masses could move to the walled cities we built to protect ourselves was outrageous! We spent our own money to protect our golf courses and land areas on high ground, ran drainage tunnels underneath, build earthquake-proof towers, and prepared for climate change flooding. Our protections were just for ourselves, not for everyone else. Those areas should have been ours and ours alone!" complained Richey, reaching for another caviar appetizer.

"Yeah, walling off areas and putting in water pumps was our own business. It really galls me!" moaned Yates as he flicked a bit of dust off his precisely ironed lapel.

"All the while spreading propaganda that climate change didn't exist!" laughed Richey. "We were good, weren't we?"

"People always believe beautiful women like me," added Celestine. "Don't they? Don't you believe me when I say there is no climate change?"

"Just about, Celestine. Now we have the last laugh! The new planet will be all ours!" added Sterling.

"If only this ride in space could hurry up. It will be an uncomfortable six more years to our new home. The court case decisions giving our beautiful towers to the masses was terrible. We had to do something to avoid jail. A new planet is a perfect answer, once we get there," stated Slone.

"We have enough Botox with we should arrive looking as young as the day we left," sighed Belle. "Smart of me to bring my esthetician."

"The new planet will be all ours for the taking!" exclaimed Joe, who didn't feel a bit guilty for helping those left behind on Earth. At least, he thought he had helped them. Some of the messages he had been getting lately were confusing, but he thought it was for the betterment of mankind.

"Well, not exactly *yours*, Joe. We will own everything. You work for us!" Kingsley dictated.

"Right, Boss, I remember," sighed Joe. "I forget I'm a menial worker when I'm constantly around all you billionaires. Say, we have a message coming in from Earth. I will check the computer and share the information in a minute."

"Message from Earth! Who cares! Soon we won't care at all. Earth is just nothing but trouble and bad news," laughed Yates.

"Boss, it is great news! The teenagers back on Earth from Green City 8763 and a couple of homeschooled kids from Northern Ireland came up with a solution for climate change, and Earth has a lot more oxygen! I'm so happy!" smiled Joe, forgetting how he daring he had been helping the common folk.

"What!" sputtered Kingsley. "What! You call that good news! Here we are in space six years from our new home when everything back on Earth is going to be OK!"

"Oh, sorry, Boss. Rotten news, right?"Joe had forgotten to cover his message tracks with Kingsley and hoped it would not cause too many problems for him later.

"Kids did that?" asked Chase in a loud voice. "Kids?"

"We've been had!" shouted Sterling.

"We could have just stayed on Earth!" moaned Yates.

"There is nothing we can do now," sighed Sloan.

They all sadly turned to view the stars and space visible through the spacecraft's window. Each person was thinking of people and lives they left behind while admiring their stunning reflections in the dark glass.

SURPRISE

At the Final Celebration

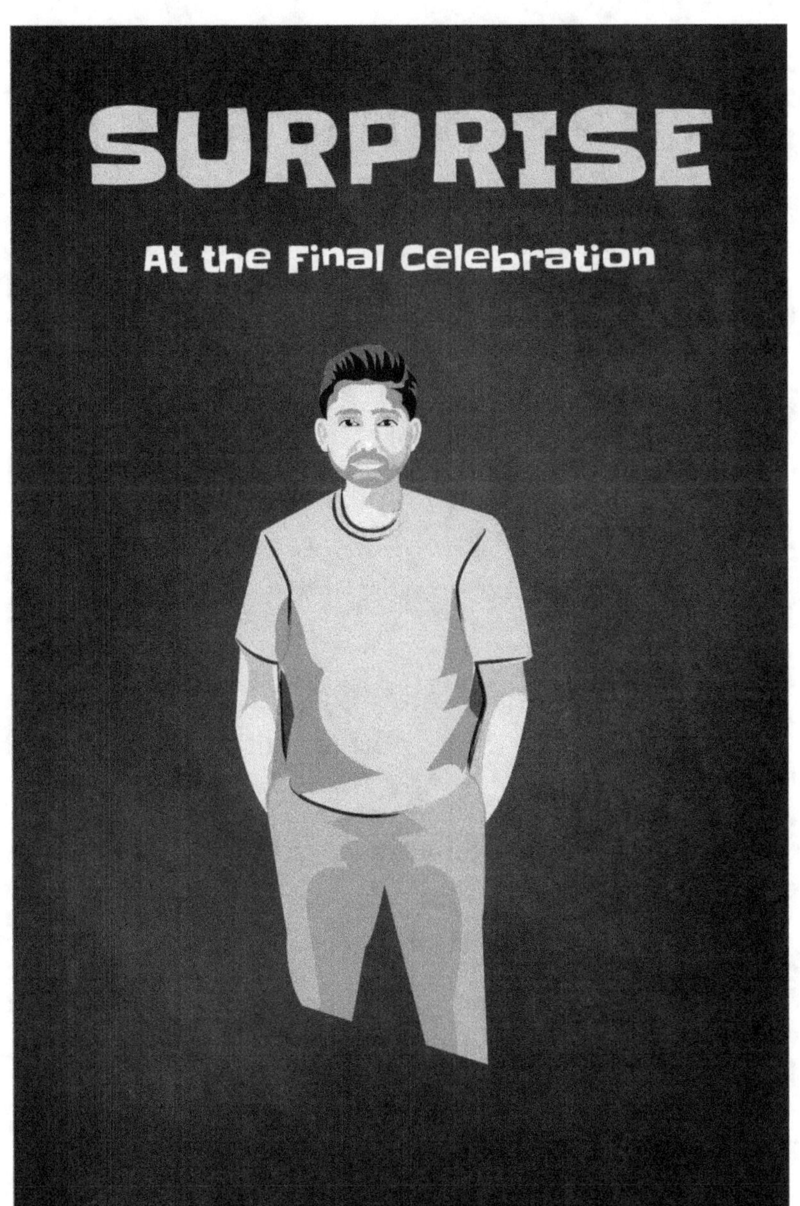

Surprise During the Final Celebration

"Well, did the message get to the *Slip Away Spaceship*?" asked Robert.

"Message received!" confirmed Tristian from the computer which communicated with the space center. The peace guards had given him Lynch's computer.

"Now they know what we achieved to start restoring Earth to its former glory," confirmed Brea.

"Ha-ha to the billionaires in space who left after creating all the havoc," said Kallista. "They got what they so *richly* deserved!"

"Agreed!" chorused Brea, Robert, and Tristian, while enjoying breathing in the fresh outdoor air and high-fiving each other.

"Good riddance to bad rubbish," laughed Tristian. "They were worse than the pollution!"

"It seems they did get a sentence, a life sentence, even though they left before the trials had even finished," mentioned Brea.

"And Earth and everyone on it didn't get a death sentence. The planet is going to recover," rejoiced Kallista.

The friends high-fived each other again.

Robert mentioned although they had learned so much about climate change solutions, they had never explored the city. The four students realized they didn't know much about the city centers, but anyway, maybe people would now be able to see different areas and do some exploring while following strict rules to prevent future global warming.

"Say," asked Robert, changing the subject. "The billionaires think they will be rich on their new uninhabited planet. Who do they think is going to buy the gold and silver they mine?"

"Well, they have a few menial workers on board. They better pay them extremely well," suggested Tristian, as the friends had the last laugh.

"What's all the yelling for?" asked Troy.

The students froze and looked at the man. They noticed Cara and Brea's grandpa enter the room, too.

Troy stared right at Robert. He had tears in his eyes.

There was a minute of deafening silence, so to speak.

"Dad?" asked Robert, running to him. "Dad?" The two hugged.

"I didn't think you would recognize me," said Troy, "but you made me very happy when you did."

"What about mom?"

Sadly, Dad shook his head and shrugged his shoulders. "I don't know, but now that things are improving, maybe we will hear from her. People have been pretty isolated. I hear now that some travel is possible, city centers are sharing lists of people looking for other people. There is always hope until we know for sure. Say, Robert, I thought of you every day."

Robert turned to Brea and demanded to know how she kept this quiet as it was obvious from their big smiles she and Cara had something to do with the reunion.

"I was just about bursting," she answered. "But it was just going to be so good if I didn't spoil the surprise!"

"Let's all go upstairs and visit your grandfather," suggested Cara. "And then you can see the new apartment Brea, her grandpa, and I will be living in a few floors up from yours.”

“Does Grandfather know? Or are we going to give him a heart attack?” wondered Robert.

The many smiles told him to wait and see what happened when they knocked on his door.

“We will come up and see the place before our hovercraft takes us to the airport,” said Tristan.

“Oh, a letter came for you,” added Cara as she turned and started up the stairs.

Tristan and Kalli silently read it together.

I can't wait to get home and tell my parents, though Kalli. All hyper-rich people are not greedy polluters! How about that! A nice philanthropist!

Wait until our parents hear this, thought Tristan. And they were so worried about paying for all of our future education! Woo hoo!

About the Author

 Carolyn Wilhelm has a BS in Elementary Education, an MS in Special Studies of Gifted Children, and an MA in Curriculum and Instruction K-12. She was a National Board Certification Middle Childhood Generalist 2004-2014. She is also a licensed, certified teacher in Minnesota through 2021. Retired, she now volunteers at an elementary school. Carolyn is a wife, mom, and grandmother. One of her now-adult children was adopted from South Korea.

After she wrote this book, in August of 2019, Carolyn completed a Climate Reality Leadership Corps training.
Recently, Carolyn and her daughter Betsy wrote *A Mom: What is an Adoptive Mother?*
Her other self-published children's picture books include the following:
Alex Asks About Auntie's Airplane Day: An Adoption Day Story
Super Spoons to the Rescue; A Math Measuring Story
The Frogs Buy A New House: An Economics Story for Children
She is the author of the Wise Owl Factory which has (mostly) free educational materials for parents, teachers, homeschools, libraries, and scout groups to use. She lives with her witty and helpful husband of 50 years and visits her children and grandchildren as often as she can.

She lives in Minnesota, and when not enjoying time with her grandchildren and creating lessons, she is involved in local writing and blogging groups — and loves taking time out in nature to observe her favorite bird, owls.

Carolyn Wilhelm
Wise Owl Factory LLC
Minnesota
https://www.thewiseowlfactory.com/
@WiseOwlFactory

About the Illustrator

Pieter Els has 30+ years of experience in facilitating courses to junior and senior facilitators, learning aids development, course material development, and marketing and web design. At first, he has earned a Graphic Design Diploma and during his career, as well as several merit bonuses for outstanding educational services and products. Later in his career he decided to also get his BTech Degree in Graphic Design. Some of the other qualifications he obtained were: Educational Technology (EDTECH) Facilitator, Learning Aids Developer, Learning Material Developer and Assessor. He also **took** several courses in Computer Based Training, E-learning Software and **attended** Middle and Senior Management Courses. One of his major responsibilities later in his career was the research on distance learning.

Pieter has an online resource company for education clip art, illustrations, tutorials and articles. His company goal is to offer the world of education more high-quality art and learning materials. Surfer Kids Clip Art was established on Christmas Day, December 25, 2013. He is happily married to Elizabe with a daughter Nika and son Eswan.

Pieter Els, Surfer Kids Clip Art

Jeffreys   Bay,   South   Africa
www.surferkiddies.com

surferkidsclipart@gmail.com

Carolyn Wilhelm

Phrases Translations

Norn Iron Spake (Northern Ireland Speech)
cry my lamps out means cry my eyes out
ascared means scared
Bout ye! means greetings
poke means ice cream
Da means father
eejit means idiot

Minnesota slang
You betcha! means you bet or yes
Dontcha know means don't you know

Secret Code

A = a
B = bub
C = cook
D = dud
E = e
F = fuf
G = gug
H = hat
I = i
J = jug
K = kook
L = lul
M = moo
N = no
O =o or oh
P = pup
Q = qut
R = rur
S = sus
T = tut
U = u
V = vuv
W = wow
X =x
Y = yuk

Z = zuz

Sample Messages to Practice the Code

Wow o wow dud i dud yuk o u sus e e tut hat a tut?

(Wow did you see that?)

I fuf o u no dud o u tut tut hat a tut tut hat e tut e a cook hat e rur sus kook no o wow!

(I found out that the teachers know)

Message 1

Pupatuttutenotutionotutotuthatisusmooes
ussusagugetuthatetuteacookhaterursuskoo
knoowowourursusecookruretutpuplulanod
udnootutwoworurruryuk

Message 1 decoded

Pay attention to this message the teachers know our secret plan and do not worry

Reply 1

wowowowwowhatokooknoewow

Reply 1 decoded

wow o wow, who knew

Reply 2

suspected as much

Reply 2 decoded

Susususpupecooktutedudasusmooucookhat

Reply 3
my dad thought so
Reply 3 decoded
mooyukdudadudtuthatougughatsuso
Message 2 from Kalli to the other three
yukoububecookarurefufulul
Message 2 decoded
you be careful
3 replies
ohhatnooh
3 replies decoded
oh no
4 mysterious messages
Hatahata
4 mysterious messages decoded
haha

Endnotes

[i] Murawski, Darlyne A., et al. "Why Insect Populations Are Plummeting-and Why It Matters." *National Geographic*, 14 Feb. 2019, www.nationalgeographic.com/animals/2019/02/why-insect-populations-are-plummeting-and-why-it-matters/.

[ii] Writers, Staff. "Effects of Climate Change on Birds." *Mass Audubon*, 2019, www.massaudubon.org/our-conservation-work/climate-change/effects-of-climate-change/on-birds.

[iii] Clark Patterson. "Urban Jungle." *The Washington Post*, WP Company, 2010, www.washingtonpost.com/wp-srv/special/metro/urban-jungle/pages/100803.html.

[iv] Wagner, Neil. "Thanksgiving Changes, Thanks to Climate Change." *HuffPost*, HuffPost, 7 Dec. 2017, www.huffpost.com/entry/climate-change-food-prices_b_1104592.

[v] Paliwal, Ankur. "A Warming Climate Could Make Pigs Produce Less Meat." *Scientific American*, 24 Sept. 2018, www.scientificamerican.com/article/a-warming-climate-could-make-pigs-produce-less-meat/.

[vi] Clark Patterson. "Urban Jungle." *The Washington Post*, WP Company, 2010, www.washingtonpost.com/wp-srv/special/metro/urban-jungle/pages/100803.html.

[vii] Lardieri, Alexia. "Trump Resort in Ireland Will Build Seawalls to Protect Against Climate Change." *U.S. News & World*

Report, U.S. News & World Report, 22 Dec. 2017, www.usnews.com/news/world/articles/2017-12-22/trump-resort-in-ireland-will-build-seawalls-to-protect-against-climate-change.

viii "Toxicodendron Radicans." Wikipedia, Wikimedia Foundation, 9 Aug. 2019, en.wikipedia.org/wiki/Toxicodendron_radicans.

ix Climate Change, United Nations. "Climate Change Threatens National Security Says Pentagon." *UNFCCC*, 14 Oct. 2014, unfccc.int/news/climate-change-threatens-national-security-says-pentagon.

x Mentaschi, Lorenzo, et al. "Global Long-Term Observations of Coastal Erosion and Accretion." *Nature News*, Nature Publishing Group, 27 Aug. 2018, www.nature.com/articles/s41598-018-30904-w.

xi Grabianowski, Ed. "How Car Crushers Work." *HowStuffWorks*, HowStuffWorks, 30 June 2011, auto.howstuffworks.com/car-crusher4.htm.

xii Du, Susan. "Minnesota's Climate Begins Its Descent toward an Unrecognizable Future." *City Pages*, City Pages, 20 Feb. 2019, www.citypages.com/news/minnesotas-climate-begins-its-descent-toward-an-unrecognizable-future/506067291.

xiii Bjorhus, Jennifer. "U Scientists: Minnesota Is One of the Nation's Fastest-Warming States." *Star Tribune*, Star Tribune, 16 Jan. 2019, www.startribune.com/u-scientists-minnesota-is-one-of-the-nation-s-fastest-warming-states/504398862/.

xiv "Loons Likely to Disappear from Minnesota Due to Climate Change, New Report Warns." *Phys.org*, https://phys.org/news/2019-10-loons-minnesota-due-climate.html.

xv Goodell, Jeff. "Forecasting Denial: Why Are TV Weathercasters Ignoring Climate Change?" *Rolling Stone*, 25 June 2018, www.rollingstone.com/politics/politics-news/forecasting-denial-why-are-tv-weathercasters-ignoring-climate-change-116394/.

xvi Marxen, Aliya. "North Shore Summer and Winter Solstice Celebrations." *Cascade Vacation Rentals*, 2019, www.cascadevacationrentals.com/summer-and-winter-solstice-celebrations.htm.

xvii Wikisource contributors. "We grow accustomed to the Dark —." *Wikisource* . Wikisource , 1 Mar. 2013. Web. 20 Jul. 2019.

xviii Rubin, Courtney. "Are You Ready for Drive-Thru Botox?" *The New York Times*, The New York Times, 30 Apr. 2019, www.nytimes.com/2019/04/30/style/botox-beauty-bars.html.

xix PBS, NH. "Bogs, Fens and Pocosins - NatureWorks." *New Hampshire PBS*, 2019, nhpbs.org/natureworks/nwep7f.htm.

xx National Geographic Society, Sue-Carly. "Peat: The Forgotten Fossil Fuel." *National Geographic Society*, 9 Nov. 2012, www.nationalgeographic.org/media/peat-forgotten-fuel/.

xxi Steer, Andrew D, and Peter Bakker. "We Help Procurement Managers Make Informed Choices about Wood and Paper-Based Products." *Home | Sustainable Forest Products*, 2019, sustainableforestproducts.org/Sustainable_Forest_Management.

xxii Oder, Tom. "16 Plants That Repel Unwanted Insects." *MNN*, Mother Nature Network, 12 July 2019, www.mnn.com/your-

[xxiii] home/organic-farming-gardening/stories/12-plants-that-repel-unwanted-insects.

[xxiii] Kenyon, Georgina. "Future - How Weeds Help Fight Climate Change." *BBC*, BBC, 10 May 2019, www.bbc.com/future/story/20190507-weeds-a-surprising-way-to-fight-climate-change.

[xxiv] Collomb, Jean-Daniel. "The Ideology of Climate Change Denial in the United States." *European Journal of American Studies*, European Association for American Studies, 2 Jan. 2014, journals.openedition.org/ejas/10305.

[xxv] "Causes." *NASA*, NASA, 2014, climate.nasa.gov/causes/.

[xxvi] King, Ledyard. "Do Wind Farms Cause Cancer? Some Claims Trump Made about the Industry Are Just Hot Air." *USA Today*, Gannett Satellite Information Network, 4 Apr. 2019, www.usatoday.com/story/news/politics/2019/04/03/cancer-causing-wind-turbines-president-donald-trump-claim-blown-away/3352175002/.

[xxvii] Worrall, Eric. "New Statesman: Rich People Plan to Flee the Earth to Escape Climate Catastrophe." *Watts Up With That?*, 21 Mar. 2019, wattsupwiththat.com/2019/03/20/new-statesman-rich-people-plan-to-flee-the-earth-to-escape-climate-catastrophe/.

[xxviii] "2009 Red River Flood." *Wikipedia*, Wikimedia Foundation, 8 July 2019, en.wikipedia.org/wiki/2009_Red_River_flood.

[xxix] "Are Category 4 and 5 Hurricanes Increasing in Number?" *Weather Underground (10.226.231.67)*, 2006, www.wunderground.com/education/webster.asp.

xxx Cappucci, Matthew, and Jason Samenow. "Historic 'Bomb Cyclone' Sets off Severe Storms, Flooding, and a 'Dangerous' Blizzard in Plains and Midwest." *The Washington Post*, WP Company, 13 Mar. 2019, www.washingtonpost.com/weather/2019/03/13/historic-bomb-cyclone-sets-off-severe-storms-flooding-dangerous-blizzard-plains-midwest.

xxxi Pappas, Stephanie. "After Death: 8 Burial Alternatives That Are Going Mainstream." *LiveScience*, Purch, 9 Sept. 2011, www.livescience.com/15980-death-8-burial-alternatives.html.

xxxii Morgan, Sam. "Daily Emissions of Cruise Ships Same as One Million Cars." *Euractiv.com*, EURACTIV.com, 10 July 2017, www.euractiv.com/section/air-pollution/news/daily-emissions-of-cruise-ships-same-as-one-million-cars/.

xxxiii Mosedale, Mike. "How Climate Change Will Impact Minnesota." *Mpls.St.Paul Magazine*, 8 July 2019, mspmag.com/arts-and-culture/climate-change-minnesota/.

xxxiv "Emerald Ash Borer." *Emerald Ash Borer - The Arbor Day Foundation*, 2019, www.arborday.org/trees/health/pests/emerald-ash-borer.cfm.

xxxv Somerville, Madeleine. "4 Brands Doing Clothes Recycling Right." *Earth911.Com*, 18 Oct. 2016, earth911.com/business-policy/clothes-recycling-4-brands/.

xxxvi Ellsmoor, James. "Trump Administration Rebrands Fossil Fuels As 'Molecules Of U.S. Freedom.'" *Forbes*, Forbes Magazine, 31 May 2019, www.forbes.com/sites/jamesellsmoor/2019/05/30/trump-administration-rebrands-carbon-dioxide-as-molecules-of-u-s-freedom/#24a5ffdb3a24.

xxxvii Harrabin, Roger. "'Restore UK Bogs' to Tackle Climate Change." *BBC News*, BBC, 22 July 2019, www.bbc.com/news/uk-49074872.

xxxviii Kepler, Dennis, and Michael Gary. "Forest Management Section - Minnesota DNR." *Minnesota Department of Natural Resources*, DNR Minnesota, 2019, www.dnr.state.mn.us/forestry/forest_management.html.

xxxix Nace, Trevor. "New Filipino Law Requires Every Student To Plant 10 Trees If They Want To Graduate." *Forbes*, Forbes Magazine, 29 May 2019, www.forbes.com/sites/trevornace/2019/05/29/new-filipino-law-requires-every-student-to-plant-10-trees-if-they-want-to-graduate/#40aa08185aeb.

xl Childress, Donna. "Tree Thinning 101." *American Forest Foundation*, 2014, www.forestfoundation.org/woodland-tree-thinning-101.

xli Dell'Amrore, Christine. "Who Were the Ancient Bog Mummies? Surprising New Clues." *National Geographic*, National Geographic Society, 18 July 2014, news.nationalgeographic.com/news/2014/07/140718-bog-bodies-denmark-archaeology-science-iron-age/.

xlii Kadri, Moin. "New Technology Can Help Reduce Risks from Landslides." *HeraldNet.com*, HeraldNet.com, 24 Apr. 2015, www.heraldnet.com/opinion/new-technology-can-help-reduce-risks-from-landslides/.

xliii Khullar, Bhavya. "Nanomaterials Could Combat Climate Change and Reduce Pollution." *Scientific American*, Scientific American, 4 Sept. 2017, www.scientificamerican.com/article/nanomaterials-could-combat-climate-change-and-reduce-pollution/

xliv Ravilious, Kate. "Terrawatch: Rocks Could Have a Role in Combatting Climate Change." *The Guardian*, Guardian News and Media, 1 May 2018, www.theguardian.com/science/2018/may/01/terrawatch-rocks-could-have-a-role-in-combatting-climate-change.

www.ingramcontent.com/pod-product-compliance
Lightning Source LLC
Chambersburg PA
CBHW060511300726

48975CB00008B/2734